Biogas Technology
in Nigeria

Biogas Technology in Nigeria

Isaac Nathaniel Itodo
Eli Jidere Bala
Abubakar Sani Sambo

CRC Press
Taylor & Francis Group
Boca Raton London New York

CRC Press is an imprint of the
Taylor & Francis Group, an **informa** business

First edition published 2022
by CRC Press
6000 Broken Sound Parkway NW, Suite 300, Boca Raton, FL 33487-2742

and by CRC Press
2 Park Square, Milton Park, Abingdon, Oxon, OX14 4RN

© 2022 Isaac Nathaniel Itodo, Eli Jidere Bala and Abubakar Sani Sambo

CRC Press is an imprint of Taylor & Francis Group, LLC

ISBN: 978-1-032-14956-1 (hbk)
ISBN: 978-1-032-14958-5 (pbk)
ISBN: 978-1-003-24195-9 (ebk)

DOI: 10.1201/9781003241959

Typeset in Times
by codeMantra

Contents

List of figures

List of tables

Foreword

COVID-19! Climate Change! Sustainability! Population Growth Rate! You might wonder why I start with those key words as an introduction to a book about biogas technology. The answer is: There are no stand-alone solutions. Everything is connected and interrelated and I encourage you, while looking at the details, to never lose sight of the grand scheme of things. Or simply put: It is complicated.

The world is facing unprecedented challenges, and immediate action is required to achieve massive decarbonization in line with keeping the rise in global temperatures below 2°C. Simultaneously, the growing global population comes with a rapidly increasing demand for energy. This increase cannot be met with higher consumption of fossil fuels, or it would lead to higher greenhouse gas emissions, which ultimately would contribute to global warming and climate change.

Governments all over the world have to urgently address these challenges and this is especially true for decision makers in developing and emerging countries, where the number of people without access to electricity remains high. Enhancing energy security while mitigating climate change is – next to securing the provision of clean water and health care – one of the biggest tasks ahead. It is essential for governments to create an enabling policy framework to encourage private-sector involvement, as these transitions will require significant investments.

"Every challenge is an opportunity in disguise", this optimistic quote by John Adams should motivate all of us and especially the readers of this book to work towards feasible solutions. And this book can be one stepping stone for such a change for improvement. Renewable energy and energy efficiency have to be the main drivers in this transformation towards a low carbon economy. This is not an easy undertaking but it is possible. More importantly, this large-scale transformation of the global energy sector is absolutely necessary.

Currently, the main renewable energy activities are led by solar, wind, and hydropower projects, yet the production of biogas has to play a key role in the diversification of energy sources. Biogas not only has a large potential as a clean and reliable energy source but can also reduce the water pollution from livestock waste, and in addition to that, the material remaining after the digestion process can be used as fertilizer.

Since 2020 the world has been dealing with a global pandemic, and it is predicted that such scenarios will occur more frequently if the human race does not introduce a more sustainable way of living. Producing enough food for an ever-growing global population while maintaining an intact environment and ensuring water and energy supply to everybody are the main challenges ahead.

Knowledge is the driving factor towards the introduction of better and sustainable technologies. The more we learn about the complexity of the situation, the better we are prepared to address those obstacles and challenges. One step, or in this case, one biogas plant, at a time.

Sabine Sibler (MSc)
Virginia Tech, USA

Preface

This book is a reflection of my 31 years' experience in teaching and research of biogas technology in Nigeria universities. This book is intended to be a teaching, research and reference material for students, researchers, professionals and policy makers working on biogas development and extension. Biogas is beginning to receive special attention as a biofuel of significance in Nigeria and the world. Politicians, farmers, policy makers, environmentalists and energy practitioners are talking about biogas.

This book is a practical guide to designing, constructing, operating and maintaining biogas plants. The book, hopefully, will correct some of the wrongly held views and opinion on biogas technology and advance its suitability as an appropriate renewable energy for use in Nigeria.

This 13-chapter book addresses important issues in biogas technology such as the economic analysis and user evaluation of biogas plants that deal with profitability indicators such as payback period and return on investment. The book also discusses key issues such as policies and standards, and utilization of biogas for electricity generation and as automotive fuel. There are chapters on the obstacles and way forward in promoting biogas in Nigeria and a curriculum for biogas education in Nigeria.

The emphasis on Nigeria in this book is to encourage and support the adoption and integration of this simple, affordable energy resource in the energy matrix to improve the energy per capita, improve living standard and stimulate college industries in Africa's most populous nation.

The co-authors, Professor Abubakar Sani Sambo and Professor Eli Jidere Bala, are big advocates of biogas technology in Nigeria and have been at the forefront of promoting the technology as teacher, researcher and Director General of the Energy Commission of Nigeria. Their experiences are captured in this book.

I trust and hope that users of this book will find it a useful literature.
Engr. Professor Isaac N. Itodo, FNSE, FAEng., FSESN, FNIAE
June, 2021

Preface

Acknowledgement

I will ever be grateful to Professor F. O. Aboaba, a retired Professor of Agricultural Engineering, University of Ibadan, Ibadan, Nigeria. He introduced me to biogas technology while teaching my Master's degree class of 1988 the course 'Power Systems'.

I am indebted to my PhD Supervisor, Engr. Professor Emmanuel Babajide Lucas, a retired Professor of Agricultural Engineering, University of Ibadan, Ibadan, Nigeria. I deeply appreciate the mentorship of Engr. Professor Abubakar Sani Sambo, an outstanding professor of renewable energy and a national merit award winner who is also a co-author of this book.

Special thanks to Mrs. Nina Ani, Managing Director and Chief Executive Officer of Avenam Links International Limited, Lagos for the pictures of balloon storage of biogas. My gratitude also goes to Rv. Fr. Dr. Godfrey Nzamujo, Founder of the Songhar Centre, Porto Novo, Republic of Benin for allowing me to take a picture of the biogas storage tank on his farm during my visit in August, 2008. I thank Sabine Sibler, Environmental Engineer, Virginia Tech, USA, for writing the foreword of this book.

Students are and will always be the donkeys of research. I recognize the hard work of my final year, postgraduate diploma, master and doctorate degree students who worked on biogas energy. Their works were a major source of reference in this book. I am grateful to them; I acknowledge Cletus C. Onuh, Benedict B. Ogar, Julius Abongwa, J. C. Uzochukwu, Jeleel A. Salaudeen, Watson Pessu, Alfred O. Okachi, John Idikwu, Joseph T. Songu, Caleb Oryiman, J. A. Onwunebe, Emmanuel Okoh, Timothy T. Wuese, Jeremiah I. Waihyo, Joseph I. Arira, Pever Enngo-Usaka, Gani E. Agyo, Genesis Ishaya, Mathew Bausa, Dr. John O. Awulu, Dr. Theresa K. Kaankuka, Paul Y. Onalo and Dorcas Kantiok.

The lockdown during the Covid-19 pandemic gave the impetus to writing this book. I am grateful to Almighty God for his protection and good health in this period. I deeply appreciate the encouragement and support of my wife and daughter.

Professor Isaac N. Itodo
June, 2021

Authors

Isaac Nathaniel Itodo is a Professor of Agricultural Engineering. He obtained a BEng from the University of Jos in 1986, and an MSc and a PhD from the University of Ibadan in 1989 and 1993, respectively. He was Head of Department (1996–2000); Dean, College of Engineering (2004–2008); and Director of Linkages (2014–2019) at the University of Agriculture, Makurdi, where he has taught for 34 years, becoming a Professor in 2003. He was President of Solar Energy Society of Nigeria (2015–2019) and a Fellow of the Nigerian Society of Engineers and the Nigerian Academy of Engineering. He has several publications in renewable energy

Eli Jidere Bala is a Professor of Mechanical Engineering. He obtained a BEng and MEng from Ahmadu Bello University, Zaria in 1977 and 1980, respectively, and has taught in the same university since 1978 where he was Head of Department (1991–1993; 1997–1999). He got a PhD from Cranfield Institute of Technology, UK in 1984. He was Rector of Abubakar Tatari Ali Polytechnic, Bauchi (1993–1997), and is currently the Director General of the Energy Commission of Nigeria, Abuja. He is a Fellow of the Nigerian Society of Engineers and the Nigerian Academy of Engineering. He has several publications in renewable energy.

Abubakar Sani Sambo is a Professor of Mechanical Engineering. He obtained a First-Class Honours degree in Mechanical Engineering from Ahmadu Bello University, Zaria in 1979 and a DPhil degree from the University of Sussex, UK in 1983. He was Director of Sokoto Energy Research Centre, Usmanu Danfodiyo University, Sokoto, and Vice Chancellor of Abubakar Tafawa Balewa University, Bauchi and Kaduna State University, Kaduna. He was also the Director General of the Energy Commission of Nigeria (2005–2012). He is a Fellow of the Nigerian Society of Engineers, Nigerian Academy of Engineering and Nigerian Academy of Science. He is a recipient of two Nigerian national honours: the National Productivity Order of Merit (1997) and the Officer of the Order of the Niger (2000). He has several publications in renewable energy.

Biogas as a Renewable Energy

1

1.1 RENEWABLE ENERGY IN NIGERIA

Renewable energy (RE), commonly referred to as non-conventional energy, is an inexhaustible energy that derives from the sun. The earth receives 95 billion MW of energy from the sun, which is about 95 million typical sized nuclear power plants. On a clear day, the insolation reaching the earth is about 1000 W/m². This value is influenced by atmospheric conditions, the earth's position in relation to the sun, and obstruction at the location. The justification for the increasing use of RE is the increase in consumption of crude oil compared to production, which at current production will result in its complete depletion in a few years.

The use of RE is driven by governments of the world, providing favourable policies to accelerate the dispersal and deployment of RE technologies (RETs). The most successful government policies on RE include feed-in-tariff, tax incentives/rebate, tax relief, RE credits/carbon credits and interest-free loans and financing. The production of energy from renewable sources is increasing. The common types of RE in use include solar, which can be thermal or photovoltaic, wind, and biofuels: bioethanol, biodiesel and biogas.

Nigeria currently distributes about 3,000 MW of electricity for consumers which is far less than the estimated 36,000 MW required. Nigeria thus has a huge electricity deficit that cannot be met from the existing hydroelectricity sources. Nigeria's electricity consumption of 18,140,000 MWh/year (12.2 W/capita) is ranked 71st in the world, which is only better than Chad (178th), Equatorial Guinea (179th), Guinea Bissau (184th), and Sierra Leon (185th) and lower than Algeria, Morocco and South Africa. The electricity

DOI: 10.1201/9781003241959-1

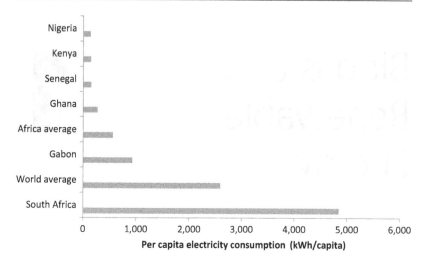

FIGURE 1.1   Per capita electricity consumption of some countries compared to Nigeria.

consumption per capita in Nigeria is 135 kWh/capita, which compares poorly with the world and Africa's averages (Figure 1.1). Nigeria currently generates a theoretical 3,900 MW of electricity while the demand is estimated at over 36,000 MW. There is, thus, a shortage of over 30,000 MW that can never be met from the existing sources. Also, the problem of transmitting grid-generated electricity to the rural areas through very difficult terrains makes electricity supply to the rural communities difficult.

The installed capacity of biogas plants, solar electricity, solar water pumping, bioethanol and small hydro-power in Nigeria as of 2017 is 586 m³, 11.16 MW, 1.05 MW, 15.3 million L/year and 37 MW electricity, respectively (Table 1.1).

1.2 BIOMASS ENERGY

Energy is the ability to do work. Work means offering services such as lighting, cooking, heating, cooling and transporting. Energy consumption is a measure of a nation's index of development and standard of living. It is estimated that in the world today about 1.6 billion people have no access to energy and about 2.6 billion rely on biomass energy sources for cooking. Energy from agricultural wastes is commonly referred to as bio-energy because it derives

TABLE 1.1   Installed capacities of renewable energy technologies in Nigeria as in 2017

RENEWABLE ENERGY SOURCE	INSTALLED CAPACITY	NO. OF INSTALLATIONS	ESTIMATED COST PER W_P OF INSTALLED SYSTEM (N)
Solar electricity for lighting	11.16 MW (excluding private installations)	-	17,000.00
Solar water pumping	1.05 MW (excluding private installations)	888	2,314.00
Biogas	586 m³ (including private plants)	25 plants	-
Bioethanol	15.3 million litres per year	-	-
Small hydro-power	37 MW electricity	-	-

Source: Itodo et al. (2017).

from organic materials (biomass). Typical examples of bio-energies, which are also called biofuels, are biogas, bioethanol and biodiesel. Biomass is the most renewable source of energy and is most widely used within the energy system. In Nigeria, the use of biomass for energy has majorly been for cooking and heating and not for electricity generation. The livestock population of 289.7 million generates about 61 million tonnes of waste in a year, 50% of which can generate 40 MW of electricity in a year when subjected to biological gasification. This amount of electricity represents 0.2% of Nigeria's electricity needs. According to the Energy Commission of Nigeria's projection, biomass will add 3,345 MW of electricity by 2025, although Nigeria presently has no single biomass-fired electricity-generating plant despite the abundance of biomass.

1.3 BIOFUELS

Biofuels are solid, liquid and gaseous fuels derived from organic matter. The common liquid biofuels in use are bioethanol and biodiesel, while the most popular gaseous biofuel is biogas. The global biofuel production increased by 10 billion litres in 2018 to reach a record of 154 billion litres. The USA, China and Brazil are the largest biofuel producers in the world. Biofuel output is anticipated to increase by 25% in the period between 2019 and 2024 to reach 190 billion litres.

TABLE 1.2 Energy demand in Nigeria, breakdown by sector

SOURCE	2010	2030	REMAP 2030
Oil, PJ	341.0	1,941.0	1,619.0
Bio-gasoline, PJ	-	-	194.0
Biodiesel, PJ	-	-	132.0
Share of RE fuels, electricity	0%	0%	17%

Source: IRENA (2018).

Currently, biofuel production and its use in Nigeria is almost zero except for a few biogas plants. There is no commercial production of bioethanol and biodiesel. Forecast shows that there will be an increasing role of biofuels as anticipated due to the demand of the various energy sources (Table 1.2). The demand for RE fuels and electricity will increase from 0% to 17% in 2030. It is expected that the biofuels share of RE in transport in 2024 will be about 90%.

1.4 BIOGAS TECHNOLOGY

Biogas is a methane-rich gas that is produced from the anaerobic digestion of cellulosic matter. The composition and properties of biogas are provided in Tables 1.3 and 1.4, respectively. The main interest in biogas is from Asia and the Pacific region. Biogas has had very little impact in Latin America. The West sees biogas technology as appropriate, while the Developing Countries think that it is a second-class technology. Biogas is used in Tanzania, Burundi, Cameroon, Benin Republic and Nigeria. About 150 mtoe of biogas will be produced globally by 2040, over 40% of which is in China and India.

TABLE 1.3 Composition of biogas

CONSTITUENT	COMPOSITION (%)
Methane	55–65
Carbon dioxide	25–45
Oxygen	0.1
Carbon monoxide	0.1
Hydrogen	1–10
Nitrogen	1–3
Hydrogen sulphide	Trace

TABLE 1.4 Properties of biogas

Property	VALUE
Heating value	22.0 MJ/m³ (15.6 MJ/kg)
Density	1,200 kg/m³ at atm. pressure
Combustion speed	40 cm/s
Air requirement	5.7 m³/m³ of air

1.4.1 Why Biogas?

Biogas technology is becoming increasingly important and acceptable for the following reasons:

1. The increasing reality of the Hubbert Peak has led to an increase in unit price of fossil fuel across the globe. The Hubbert Peak is the point of maximum production of oil. The point half of the recoverable oil that ever existed on the planet has been used. Although oil still exists, it is much harder and more expensive to recover.
2. Availability of feedstock for biogas production. Nigeria is blessed with a huge amount of feedstock for generating biogas: Nigeria produces about 61 million tonnes of animal wastes per year (Table 1.5).
3. Biogas production and use ensures a further reduction of Green House Gas (GHG) emission, thus preserving the planet. Although the contribution of Africa to GHG emissions is estimated to be about 3%, it is important to keep it as low as possible by using RE sources like biogas. Biogas saves on carbon emission by preventing the escape of methane – a strong GHG with global warming

TABLE 1.5 Calculated manure production of Nigeria's livestock

LIVESTOCK	MANURE PRODUCED (MILLION TONNES)	POPULATION BASED ON FMA, 1997 (MILLION)	MANURE PRODUCED 2001 CALCULATED (MILLION TONNES)
Cattle	170.4	21	197.6
Sheep	13	38.5	15.1
Goat	21.1	62.4	24.5
Pig	13.2	9.6	15.3
Poultry	28.1	42.9	32.6
Total	245.9		285.1

Source: ECN (2005).

potential. Biogas is carbon-neutral because the carbon dioxide produced is used up by plants, which the animals that produce the wastes feed on. It is generally accepted that biofuels are carbon-neutral because no net CO_2 is added to the atmosphere (Figure 1.2).

4. Biogas is currently an acceptable energy worldwide. Biogas for energy is a growing trend in about 50 countries, especially China, Sweden, Germany and the UK. Several European countries are expanding their total share of power from biomass – Austria (7%), Finland (20%) and Germany (5%). Biogas is the key to China's rural development. In 2007, 26.5 million biogas plants with a combined output of 10.5 million m^3 biogas and household digesters were found throughout the country. Biogas is widely used in Asian countries such as India and China, where they have minimized the drudgery associated with the production and use of energy in households. Biogas is key to China's rural development. Sub-Saharan Africa has the local capacity for biogas technology.

5. The appropriateness of biogas technology is in its sustainability, simplicity and affordability. Sustainable biogas production will improve and upgrade the economies of developing countries as it will stimulate economic activities, which will in turn increase employment. The energy profile of Sub-Saharan Africa regions appears to be the worst and is directly related to the state of under development, a region characterized by high cost of petroleum products, very low electricity generation and very low energy per capita.

6. The technology ensures energy independence as a unit can meet the needs of a family or community, thus improved energy security resulting from decentralized electricity generation.

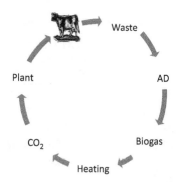

FIGURE 1.2   Carbon-neutral cycle of biogas.

7. Biogas is environmentally friendly. The production of biogas is an anaerobic treatment of wastes, which eliminates harmful pathogens, thus eliminating environmental pollution.

8. It is claimed that its value as a fertilizer could double crop yield because anaerobic treatment conserves nutrients such as N, P and K for soil fertility.

9. The treated effluent from the anaerobic digestion process is a good animal feed when treated and mixed with molasses and grains.

10. No food versus fuel controversy. Biogas, which is produced from agricultural wastes unlike bioethanol and biodiesel, does not compete for land that should be used for food crops.

1.4.2 Limitations of Biogas Technology

The disadvantages associated with biogas technology include the following:

1. Gas yield from the digester may not be steady, which therefore makes it unreliable, thereby necessitating storage.

2. The main problem associated with the use of this simple technology is that it is a low-pressure gas production system and as such cannot be bottled for use outside the site of production, thereby restricting the technology only to the site of production. However, biogas can now be compressed and used outside the point of production (Figure 1.3).

FIGURE 1.3 Storage tanks for biogas compressed from a fixed dome plant at the Songhai Farm Centre, Porto Novo, Republic of Benin.

1.4.3 The Challenges of Biogas Technology in Nigeria

The deployment of biogas technology in Nigeria is hindered by many factors, which include:

1. A lack of framework for sustainability. The lack of a framework such as the Khadi and Village and Industries Commission of India that designs, deploys and disseminates biogas plants across India.
2. Insufficient training, information and education, which has resulted in lack of technical expertise in biogas technology, have greatly slowed the deployment and dissemination of the technology. There are only about 25 biogas plants in Nigeria as of 2007, out of which a few are functional. The research and development of the technology is very low and mostly restricted to the two energy research centres at Usmanu Danfodiyo University, Sokoto and the University of Nigeria, Nsukka. Is ECN the only effective and visible promoter of RE in Nigeria? Where are the other advocates?
3. Lack of regulation. Who constructs the plants? Which agency is mandated to approve the design and supervise the construction? Which agency sanctions mis adventurers?
4. Absence of public ownership. Where are the biogas plant companies and entrepreneurs? Where are the marketers of biogas appliances? How many building engineers, town planners, farmers, artisans, health workers and students are involved in our workshop?
5. Lack of financial incentives. The recent 140 billion Naira (N140 billion) which the Central Bank of Nigeria set aside to support the private sector in the 5 million Solar Home System programme should have been extended to biogas technologies for the household and services sectors of the country.
6. Where are the standards? Any biogas standard to protect the interest of users? Any building regulations supporting/encouraging the building of biogas plants in building code?
7. Lack of awareness of the enormous potentials of bio-energy among politicians, media and the public is largely responsible for its under deployment.
8. Lack of institutional framework for promoting and sustaining bio-energies.

Planning for a Biogas Plant

2

Planning for a biogas plant requires determining the energy demand, biogas production and balancing gas production and demand. The process is illustrated in Figure 2.1.

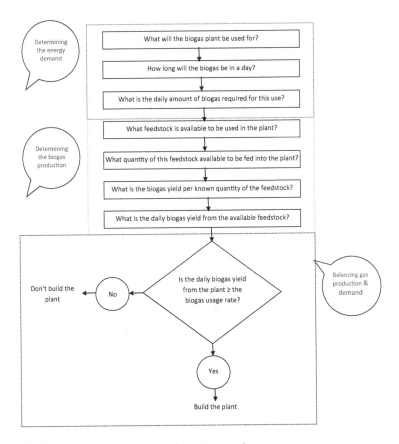

FIGURE 2.1 Process of planning for a biogas plant.

DOI: 10.1201/9781003241959-2

2.1 DETERMINING
THE ENERGY DEMAND

The energy required for use to be provided by the proposed biogas plant referred to as the usage rate is determined from:

1. How long will the biogas plant be used to provide the required energy in a day?
2. What is the daily quantity of biogas required for this use?

2.2 DETERMINING THE BIOGAS
PRODUCTION

The amount of biogas yield from the feedstock to be used in the plant is determined from:

1. What feedstock is available to be used in the plant?
2. What quantity of this feedstock is available to be fed into the plant daily?
3. What is the biogas yield per known quantity of the feedstock?
4. What is the daily biogas yield from the available feedstock?

2.3 BALANCING GAS
PRODUCTION AND DEMAND

The decision to build the plant or otherwise is made from balancing gas production and demand. This is accomplished by determining whether the daily biogas yield from the plant is greater than the biogas usage rate for the proposed use. If the daily biogas yield is greater than the quantity of biogas required for the proposed use, the plant is recommended to be built, and if it is not greater, it is not recommended to be built.

2.4 A PLANNING GUIDE

A planning guide is a checklist and fact-list of all the parameters that will enable a realistic decision to be made on whether or not to build a biogas plant. Holistically, the planning to build a biogas plant must consider the following:

1. The socio-economic profile of the user
2. Details and site characteristics of the location of the proposed plant
3. Availability of feedstock
4. Energy demand and supply and a balance between them
5. Availability of construction materials
6. Availability of gas appliances
7. Table 2.1 is the detailed planning guide for a biogas plant adapted from a study by Werner et al. (1989).

TABLE 2.1 A planning guide for biogas plants

ITEM	DATA		RATING
1.0 Addresses	1.1 Plant acronym		
	1.2 Location		
	1.3 Private or government-owned plant?		
	1.4 Name of extension officer		
	1.5 Address of extension officer's organization		
	1.6 Type of biogas plant		
2.0 Socio-economic profile of user	2.1 Family or community use?		
	2.2 Gross income per annum		
	2.3 Occupation		
	2.4 Cultural and/or religious taboos		
	2.5 Main source of energy		
	2.6 Alternate source of energy		
	Overall rating for 2.0		

(Continued)

TABLE 2.1 (Continued)   A planning guide for biogas plants

ITEM		DATA	RATING
3.0 Location and site characteristics	3.1 Mean annual ambient temperature		
	3.2 Mean annual rainfall	Annual rainfall > 1,500 mm is unfavourable	
	3.3 Mean annual wind speed		
	3.4 Soil type		
	3.5 Ground water table		
	3.6 Proximity of location to underground water sources	Should not be < 30 m	
	Overall rating for 3.0		
4.0 Energy need	4.1 Proposed use of biogas from the plant		
	4.2 Duration of use per day		
	4.3 Quantity of biogas required for use per day	Figure 2.2	
5.0 Feedstock availability (Livestock inventory)	5.1 Number of animals		
	5.2 Quantity of waste generation per animal per day	Table 2.2	
	5.3 Total quantity of waste generated per day		
	Overall rating for 5.0		
6.0 Feedstock availability (Human waste inventory)	6.1 Number of households		
	6.2 Number of persons per household		
	6.3 Waste generation per person per day		
	6.4 Total quantity of waste generated per day		
	Overall rating for 6.0		
7.0 Energy supply	7.1 Biogas yield per kg of feedstock type	Figure 2.3	
	7.2 Quantity of biogas yield per available feedstock per day		

(Continued)

TABLE 2.1 (Continued) A planning guide for biogas plants

ITEM	DATA	RATING
8.0 Balance between energy need and supply	8.1 Daily biogas yield from plant > daily biogas usage rate	
	8.2 Daily biogas yield from plant < daily biogas usage rate	
	Overall rating for 8.0	
9.0 Availability of construction materials	9.1 Bricks/blocks/ stones	
	9.2 Sand	
	9.3 Metal	
	9.4 Pipes and fittings	
	9.5 Miscellaneous	
	Overall rating for 9.0	
10.0 Availability of gas appliances	10.1 Cookers	
	10.2 Lamps	
	10.3 Hybrid carburetors	
	10.4 Generators	
	Overall rating for 10.0	
Summary of ratings	2.0	
	3.0	
	5.0/6.0	
	8.0	
	9.0	
	10.0	
	Overall rating for the project	

Ratings: P, Favourable; 0, Unfavourable but not too serious or can be corrected; N, Unsuitable.

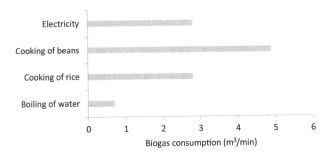

FIGURE 2.2 Quantity of biogas required for various uses.

TABLE 2.2   Fresh waste production from various animals

ANIMAL	FRESH WASTE PRODUCTION (KG/DAY)
Cattle	9.0
Pig	4.0
Poultry	0.04

Source: Itodo et al. (1999).

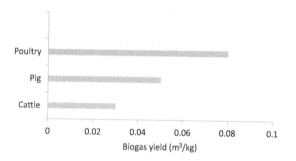

FIGURE 2.3   Biogas yield from various waste types in Nigeria.

2.5 CHECKLIST FOR BUILDING A BIOGAS PLANT

Building of the biogas plant commences after a positive rating of the parameters evaluated in the planning guide. What to do in building a biogas plant include the following:

1. A site plan of the location of the plant including all surrounding buildings, gas pipe network from the plant to the point of use, site for storage of the effluent from the plant and the relative position of the plant to underground water sources within the location.
2. A detailed technical drawing showing all plant components with dimensions: the influent and effluent pits, influent and effluent pipes, digester well, gas holder, gas pipe and slurry storage slab.
3. A bill of quantity for procurement of materials for building the plant, which should include the metal parts of the plant, tools to be used in constructing the plant, paint and sealant for the gasholder, gas appliances, skilled and unskilled labour required for building the plant.
4. A Gaunt chart of timelines for the activities, which should include the times for supervision and commissioning of the functional plant, and training of the users.

Economic Analysis and User Evaluation of Biogas Plants

3

3.1 ECONOMIC ANALYSIS OF BIOGAS PLANTS

The decision to build or not to build a biogas plant depends on both technical and economic considerations. The final decision is most often based on a cost-benefit analysis of the project. The benefits of the plant include reduced work load and work time to the rural woman in terms of the distance travelled, number of hours and energy spent in hewing wood for cooking compared to the use of a biogas plant, reliability of the fuel wood energy supply compared to that of a biogas plant and other such benefits as improved health and hygiene. Importantly, government and credit institutions such as community thrift groups are interested and will require an economic evaluation of the project before supporting the financing of a plant. The financial capacity of individuals and/ or groups desiring to build a plant is also an important economic consideration. The economic evaluation of a biogas plant involves the calculation of the payback period and the profitability. Irrespective of these determining parameters, a biogas plant is economically viable if $1\,m^3$ of biogas is produced from $1\,m^3$ of digester volume or 18–$20\,m^3$ of biogas is produced from $1\,m^3$ of feedstock.

DOI: 10.1201/9781003241959-3

3.1.1 Payback Period

The payback period is the time it takes to recover the cost of investing in an enterprise. It is the length of time it takes to attain the break-even point in the use of a biogas plant. The impetus for investing in a biogas plant is directly related to its payback period. A short payback period, a period less than 10 years indicates a positive investment. On the contrary, a payback period of 10 years or more, which may even exceed the service life of a plant is negative. The payback period is a simple way to evaluate the risk associated with investing in a biogas project. The payback period is expressed in equation 3.1.

$$t_{pb} = \frac{I_o}{R} \tag{3.1}$$

where

t_{pb} is the payback period, years

I_o is the cash outlay, which is assumed to occur entirely at the beginning of the project, currency

R is the amount of net cash flow generated by the project per year, which is assumed to be the same every year, currency

Example, if a man invests 500,000 Naira (N500,000) to build a biogas plant that has a cash flow of 100,000 Naira (N100,000) per year, then the payback period is 5 years (60 months).

3.1.2 Calculating the Payback Period for a Floating-Drum Biogas Plant

Calculate the payback period for a $4\,m^3$ floating-drum biogas plant for powering an electric generator. Table 3.1 is calculation of the payback period. The data sheet for the economic analysis is on Table 3.2. The payback period and return on investment (ROI) for the plant is 9.83 years and 7.1%, respectively. These indicators would have been much better if the effluent from the plant is sold as an organic fertilizer, thus increasing the revenue, increasing the returns and ultimately increasing the profitability of the plant.

The payback period for the plant will occur after the 9th year and before the 10th year of use, when the cumulative returns is equal to the cost of investment of 649,000 Naira (N649,000). The payback period is calculated thus:

$$t_{pb} = 9 + \frac{\left(I_o - \text{Cumulative returns just before break} - \text{even}\right)}{\text{Returns at cumulative returns after break} - \text{even}}$$

TABLE 3.1 Calculation of the payback period for a 4 m³ floating-drum biogas plant

YEAR	RETURNS (N)	CUMULATIVE PAYBACK (N)
1	66,000.00	66,000.00
2	66,000.00	132,000.00
3	66,000.00	198,000.00
4	66,000.00	264,000.00
5	66,000.00	330,000.00
6	66,000.00	396,000.00
7	66,000.00	462,000.00
8	66,000.00	528,000.00
9	66,000.00	594,000.00
10	66,000.00	660,000.00
11	66,000.00	

$$t_{pb} = 9 + \frac{(649,000 - 594,000)}{66,000} = 9 + 0.83 = 9.83 \text{ years} = 118 \text{ months}$$

The payback period for the plant is 9.83 years (118 months) for a plant with an estimated service life of 15 years (180 months).

3.1.3 Return on Investment

The profitability of a biogas plant or ROI is the ratio between net profit (NP) and cost of investment. It is used to evaluate the efficiency of an investment or to compare the efficiencies of several different investments. A high ROI indicates high profitability. The ROI of a biogas plant is calculated from equation 3.2.

$$\text{ROI} = \frac{\text{NP}}{I_A} \times 100 \tag{3.2}$$

The NP is calculated from equation 3.3.

$$\text{NP} = R - D_p \tag{3.3}$$

where
   R is the annual return that is calculated from equation 3.4

   R = Total annual income − Total annual expenditure $\tag{3.4}$

TABLE 3.2 Data sheet for economic analysis of biogas plants

Project title: A biogas plant for electricity location: Makurdi, Nigeria Owner: Mr. Raphael Enemari
Type of plant: A floating-drum continuous-flow plant digester volume: 4 m3 type of waste: Cattle waste
Estimated service life: 15 years

ITEM	0	1	2	3	4	5	6	7	8	9	10	11	12	13	14	15
													YEAR			
1.0 Investment Costs																
1.1 Planning, design, supervision consultancy	150															
1.2 Bricks	60															
1.3 Cement	90															
1.4 Sand	15															
1.5 Water	4															
1.6 Flat metal sheet	40															
1.7 Hybrid carburetor and accessories	20															
1.8 Electric generator	200															
1.9 Digging of digester well	20															
1.10 Labour	50															
1.0 Total investment cost	649															

(Continued)

TABLE 3.2 (Continued) Data sheet for economic analysis of biogas plants

Project title: A biogas plant for electricity location: Makurdi, Nigeria Owner: Mr. Raphael Enemari
Type of plant: A floating-drum continuous-flow plant digester volume: 4 m3 type of waste: Cattle waste
Estimated service life: 15 years

ITEM									YEAR							
	0	1	2	3	4	5	6	7	8	9	10	11	12	13	14	15
2.0 Income																
2.1 Energy revenue[a]		120	120	120	120	120	120	120	120	120	120	120	120	120	120	120
2.0 Total income		120	120	120	120	120	120	120	120	120	120	120	120	120	120	120
3.0 Expenditure																
3.1 Repair and maintenance		30	30	30	30	30	30	30	30	30	30	30	30	30	30	30
3.2 Transport of wastes[b]		24	24	24	24	24	24	24	24	24	24	24	24	24	24	24
3.0 Total expenditure		54	54	54	54	54	54	54	54	54	54	54	54	54	54	54
4.0 Returns (2.0 – 3.0)		66	66	66	66	66	66	66	66	66	66	66	66	66	66	66
5.0 Depreciation		43	43	43	43	43	43	43	43	43	43	43	43	43	43	43
6.0 Net profit (4.0 – 5.0)		23	23	23	23	23	23	23	23	23	23	23	23	23	23	23

The amount is in thousands of the Nigeria currency Naira.
a The amount that should have been paid as electricity tariff for use of grid electricity that is now being saved by use of the plant.
b The cost of hiring a tractor per hour once in a month for transport of waste to the site of the plant.

The depreciation (D_p) is calculated from equation 3.5

$$D_p = \frac{\text{Initial investment}}{\text{Life of the plant}} = \frac{I_o}{L_s} \qquad (3.5)$$

The life of a biogas plant is between 10 and 15 years.

The I_A is the average capital investment per time interval and is calculated from equation 3.6 (Werner, 1989).

$$I_A = \frac{\text{Initial investment}}{2} = \frac{I_o}{2} \qquad (3.6)$$

3.1.4 Calculating the Return on Investment for a Biogas Plant

Calculate the ROI for a privately owned $4\,m^3$ floating-drum biogas plant for powering an electric generator using the data on Table 3.2.

NP = N23,000.00
I_o = N649,000.00

$$I_A = \frac{649,000}{2} = N324,500.00$$

$$\text{ROI} = \frac{23,000}{324,500} \times 100\% = 7.1\%$$

3.2 USER EVALUATION OF BIOGAS

A user evaluation of a proposed biogas plant is mainly by undertaking a socio-economic evaluation of the project. A user may still proceed to building a plant even if the economic analysis of the project is not favourable because the benefits as established by a socio-economic analysis outweigh the negative economic indicators. Traditional, cultural and religious considerations sometimes influence the building of biogas plants even if the economic analysis shows a positive outcome. For example, some religious sensibilities do not allow for the rearing of certain animals. Therefore, a user will not use the waste from such animals. In such a case, the biogas plant is not feasible even with a positive economic analysis if it is the only feedstock available for use in the proposed plant. The details of a socio-economic analysis of a biogas plant are contained in Table 3.3.

TABLE 3.3 Socio-economic analysis of a biogas plant

S/NO.	SOCIO-ECONOMIC PARAMETER	APPLICABLE	POSSIBLY APPLICABLE	NOT APPLICABLE
1.0	Assured, regular energy supply			
2.0	Upgrading of women's work			
2.1	Elimination of the drudgery associated with current energy procurement tasks			
2.2	Time saved in the procurement of current energy in use			
2.3	Saving in transport cost or human haulage of the energy source in use			
3.0	Higher prestige/pride of ownership of a biogas plant			
4.0	Improved hygiene and sanitary conditions through better disposal of wastes			
5.0	Improved agricultural production by use of effluent from the plant			
6.0	Direct handling of faeces/ wastes in the loading the plant where applicable			
7.0	Traditional, cultural and religious taboos, such as the use of wastes from animals that are considered as taboo			
Total (%)				

Source: Adapted from OEKOTOP.
+, applicable; o, possibly applicable; –, not applicable.

Determining the Acceptance of Biogas Plants

4

There is currently a lot of interest in biogas technology in Nigeria and many other developing sub-Sahara African countries because of the low energy per capita and prevailing energy poverty. The decision to launch a biogas extension programme to build more biogas plants will depend on the detailed analysis of the experiences on existing plants. Thus, a robust biogas extension programme will depend on the social acceptance of the existing plants. Table 4.1 contains the factors determining the acceptance of biogas plants for establishing a biogas extension programme.

4.1 DETERMINING FACTORS OF ACCEPTANCE FOR BIOGAS PLANTS

Some of the factors that determine the acceptance of biogas plants include the role of women, functionality of existing plants, cost-benefit of the plants and finance. Others are religious beliefs and cultural taboos, agricultural practices and competent artisans.

DOI: 10.1201/9781003241959-4

TABLE 4.1   Determining factors for establishing a biogas extension programme

FACTOR	DETAILS	YES OR NO	ACCEPTABLE OR NOT ACCEPTABLE
Role of women	Are the women willing to handle animal waste?		
	Do the women have freedom to participate in group discussions planning for biogas plants?		
Functionality of existing biogas plants	Are existing plants meeting the energy need of the family?		
	Are the plant and appliances user friendly?		
	Is the repair and maintenance of the plants and appliances too frequent?		
	Are the appliances available?		
Cost-benefit of existing biogas plants	Has the use of existing plants sufficiently replaced the drudgery associated with hewing and fetching of firewood?		
	Is the time saved from use of biogas considerable?		
Financing of biogas plants	Does the family or individual have enough resources to build a biogas plant?		
	Is there an agency or organization funding the building of biogas plants as a long-term loan/subsidy?		
Religious beliefs and cultural taboos	Any religious belief and/or cultural/traditional taboos of certain animals?		

(Continued)

TABLE 4.1 (Continued) Determining factors for establishing a biogas extension programme

FACTOR	DETAILS	YES OR NO	ACCEPTABLE OR NOT ACCEPTABLE
Agricultural practices	Do their agricultural practices include animal husbandry?		
	Is the family livestock size adequate for the desired plant?		
	Is there a nearby livestock farm that can provide the required quantity of waste for the plants?		
Competence of artisans	Are there qualified craftsmen available for building, repair and maintenance of plants and appliances?		
	Are the artisans interested in biogas plants?		
	Are there training programmes for these artisans?		

4.1.1 The Role of Women

Women and children most importantly operate the biogas plants. They tend the animals, collect the wastes/droppings, fetch the water to mix with the wastes, operate the appliances, do the cooking, spread the effluent from the plant for drying and bag the effluent and carry to the farm for fertilization of the fields. Therefore, socio-cultural factors such as the willingness of the women to handle animal wastes and permission to participate in group discussions in planning for building biogas plants are major decision drivers for the biogas extension programme.

4.1.2 Functionality of Existing Biogas Plants

How well the existing plants are producing gas to meet the family's energy needs is an important determinant in the quest for building new biogas plants. The user friendliness of the plant and the appliances, repair and maintenance of the plant and appliances are also important factors.

4.1.3 Cost-Benefit of the Plant

How much has the biogas plant replaced the drudgery associated with the hewing and fetching of firewood for the energy need of the family? How much time is saved by the use of the plant and deployed to other meaningful endeavours?

4.1.4 Finance

Do families have enough resources to build a plant that can meet their daily energy needs? Is there an agency that finances the building of plants as long-term zero interest loan/subsidy?

4.1.5 Religious Beliefs and Cultural Taboos

Religious beliefs and cultural taboos such as abhorrence of certain animals and their fecal deposits and waste handling play community role in adopting the technology among certain homogeneous groups and settlements. For example, some religions and tribes abhor the rearing of certain animals.

4.1.6 Agricultural Practices

Does the agricultural practice of the family or community include rearing of animals in addition to arable crop farming? What is the size of the family livestock? This is directly related to the availability of sufficient waste that will generate the required amount of energy to meet the daily energy requirement of the family? Is there a livestock farm of many animals from which sufficient wastes can be obtained for the several proposed biogas plants within its proximity?

4.1.7 Competence of Artisans

Are there qualified craftsmen (masons, welders, plumbers, etc.) available for building, repair and maintenance of biogas plants in the village? Are the artisans interested in biogas plants? Are there training programmes for these artisans?

Production and Purification of Biogas

5

5.1 THE ANAEROBIC DIGESTION PROCESS

Biogas is produced when biomass (organic) materials are subjected to biological gasification. The organic materials are held in a digester or reactor. Animal wastes and plant residues are organic compounds that contain carbon and nitrogen in the appropriate ratio, which makes them good feedstock for the production of biogas.

The gas is produced from a three-phase process, namely, hydrolysis, acid-forming (acidogenic) and methane-forming (methanogenic) phases (Figure 5.1). It is a biological-engineering process in which a complex set of environmentally sensitive micro-organisms are involved. In the hydrolysis phase, extracellular enzymes secreted by acidogens break down the complex organic material into simple, soluble molecules. These molecules are broken down into volatile fatty acids (VFAs) (e.g., propionic and butyric acids), carbon dioxide, ammonia and hydrogen by acidogens in the acid-forming phase. In the methane-forming phase, methanogens or methane-formers convert the VFA into methane, carbon dioxide, carbon monoxide, nitrogen and hydrogen sulphide. Synthesis of carbon dioxide and hydrogen also takes place in this phase to form methane and water.

Staphylococcus species of bacteria and *fusarium* species of fungi have been identified in some slurries undergoing anaerobic digestion. These bacteria produce biogas by secreting extracellular enzymes such as lipase, protease and cellobiase that breakdown cellulose, carbohydrates, proteins

DOI: 10.1201/9781003241959-5

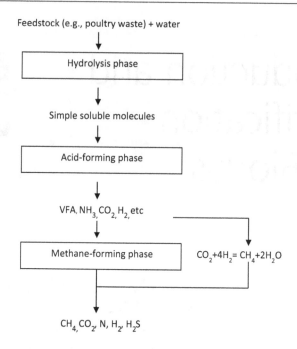

FIGURE 5.1 Biogas production process.

and lipids into simpler molecules during the hydrolysis phase of biogas production. These bacteria also enable the conversion of the hydrolyzed simple molecules into VFAs, alcohols and aldehydes and their subsequent conversion into biogas. The fungi enable the breakdown of compounds that are mainly carbohydrates into acid intermediates required for biogas production. The yeast cells enable the production of alcohols, which is a suitable substrate for the methanogens in biogas production.

5.2 FACTORS AFFECTING BIOGAS PRODUCTION

Temperature influences biogas production because it affects the activities of the micro-organisms that convert organic feedstocks into biogas. Temperature affects the secretion of enzymes by the micro-organisms. The temperature of the digester or reactor in which the organic material is being actively digested is classified as psychrophilic (below 20°C), mesophilic (20°C–40°C) and thermophilic (40°C–65°C). There are advantages and disadvantages of bioconversion at these temperature ranges. There may be no advantage in producing biogas at the psychrophilic temperature range because conversion at this temperature is slow and is usually not complete, which results in very low gas yield. A longer detention time of the slurry in the digester is required to enable a substantial conversion into biogas. Heating of the digester is required to improve its performance.

The production of biogas at the mesophilic temperature range is easily attained in the Tropics since this range of temperature corresponds to the ambient temperature. Consequently, no heating is required which makes the cost of production cheaper. However, longer detention time may be needed to ensure complete conversion of the slurry.

Digestion at the thermophilic temperatures produces higher quantity of gas, allows heavier loading rate of slurry into the digester and a lower detention time because of the higher rate of conversion. Few pathogens can survive at the thermophilic temperatures, which makes biogas production more sanitary. The slurry becomes less viscous at this temperature. This makes the pumping, transport and handling of slurries easier. The digestion at this temperature range is easily upset because of the 'shock' the temperature fluctuation causes to the micro-organisms.

The slurry in the anaerobic digester that is being acted upon by microorganisms has a solid and liquid component. The solid component is referred to as the total solid concentration (TS) of the slurry. It is the TS that contain the essential nutrients that the micro-organisms feed on to breakdown into biogas. The convertible component of the TS is the volatile solids (VS) concentration, which is usually as a rule of the thumb, estimated to be about 70% of the TS. Slurries are classified as low (less than 10%TS) and high (greater than 20%TS). Low TS slurries are easily pumped and do not block pumps, pipes and valves compared to high TS. The solid matter separates and accumulates at the bottom of the digester instead of remaining in suspension if the slurry is too thin. This ultimately leads to low gas yield. Conversely, the biogas that is produced is trapped within the digester and cannot escape to the headspace for storage or use if the slurry is too thick. This also results in low gas yield. The efficiency of anaerobic bio-conversion of wastes into biogas (AD_e) is estimated from equation 5.1.

$$AD_e = \frac{\text{Weight of VS converted to biogas}}{\text{Weight of VS available}} \times 100\% \qquad (5.1)$$

The residence time of the slurry in the digester is referred to as the detention time for a batch type digester or the hydraulic retention time for a continuous-flow digester. The retention time influences the quantity of gas produced, the size of the digester, quantity of slurry fed into the digester and how efficiently the slurry is bio-converted into biogas. During conversion of slurry into biogas, first an acid intermediate is produced. The acid intermediate is subsequently acted upon and converted into biogas. How long the slurry is held in the digester is determined by how quickly and efficiently the acid intermediate is converted into biogas. Digestion at the thermophilic temperatures generally has shorter retention times because of the higher conversion rate of the slurry, which also allows high loading rate of the slurry into the digester to quickly replace the used acid intermediates for sustained gas production. Most studies and experiences have shown that a 30-day retention time is good enough for the biodegradation of the slurry and destruction of harmful micro-organisms in the waste.

The loading rate of waste into the digester affects biogas production. The amount of waste that is fed into the digester is dependent on the hydraulic volume of the digester, how long the slurry stays in the digester and how efficiently the slurry is bio-converted into biogas. The TS concentration of the slurry is increased with higher loading rate, which causes some inhibitory compounds to gather in the digester that reduces the gas yield. Increasing the loading rate increases gas yield until an optimum value is reached after which the conversion efficiency of the system begins to decrease and gas yield also begins to decline. A loading rate of 1.5 and 5 kg/m^3/day is admissible in simple and large plants, respectively.

How well the acid intermediates of the slurry in the digester are being converted into the methane-rich biogas is easily measured from the pH of the slurry. Biogas is best produced when the pH of the digesting slurry is neutral (7.0–7.2) but gas production can occur at the pH between 6.6 and 7.6, beyond which gas production becomes less efficient. A neutral pH indicates that the acid intermediates are being used up by the methanogens that produce biogas, which means that there is a balance between the acid and methane formers (bacteria) in the slurry. If the pH is too low, it shows that the acid intermediates are not being used up as quickly as they are being produced. At the appropriate pH of 7.0 and 7.2, a steady-state condition of the digester is achieved. The acidity of the substrate can be improved (increasing the pH value) by reducing the quantity of fresh slurry fed into the digester. High acidity (low pH) in the slurry may also be due to the absence of natural alkalinity in the waste such as hydroxides, carbonates and bicarbonates of calcium and magnesium.

The micro-organisms that bio-convert wastes into biogas feed on the carbon and nitrogen content of the wastes to grow and multiply their population. These micro-organisms need nitrogen to secrete the enzymes that are required to use the carbon. If the nitrogen content in the waste is small and insufficient, the bacteria will be unable to use the carbon and if it is too much, particularly in the form of ammonia, the growth of the bacteria is inhibited. It is mainly the carbon in the waste that is consumed by the bacteria to produce biogas. The optimum carbon-to-nitrogen (C:N) ratio for the proper performance of the bacteria is between 20:1 and 30:1. Carbon-deficient slurry can be improved by adding high carbon content materials like plant wastes.

Toxicity in the waste being digested impedes biogas production. High concentration of some compounds creates a toxic environment to methanogens, thus resulting in low gas yield and sometimes digester failure. The presence of heavy metals like copper, zink and nickel in the waste are needed but are toxic at high concentrations. Also, feces of animals that have been treated with antibiotics that are used in anaerobic digestion causes very low gas yield. Generally, any waste stream with ammonia concentration greater than 3,000 mg/L has an adverse effect on biogas production. Sometimes, these heavy metals may be present in the waste but do not impact negatively on the gas production process because the waste contains some other chemicals that suppress their effect and can remove them by precipitation as sulphides. The digestion system can sometimes tolerate and become used to the presence of these toxic compounds. For example, a digester with high concentration of sulphur compounds like sulphates can admit a high concentration of heavy metals because heavy metals in the form of sulphates have no toxic effect on the gas production process. The anaerobic digestion of proteins contained in the waste in the digester results in aromatics like phenol, p-cresol, ethylphenol, indole and skatole. These aromatics may be inhibitory to the anaerobic digestion process. For example, phenol has been found to be significantly inhibitory at concentrations between 0.1% and 0.4% TS and that concentrations above 12.4% of the TS are toxic. High concentrations of VFAs are toxic to the anaerobic digestion process. The high concentrations of VFA affect methanogenic bacteria; since the substrate becomes too acidic for their activity. Their cell wall is more permeable to un-dissociated molecules in comparison with their ionized state. This means that more VFA will be taken up by the organisms at low pH.

5.3 METHODS OF IMPROVING BIOGAS PRODUCTION

Literature emphasizes that the anaerobic digestion process can be improved to enhance greater gas production by the introduction of suitable media material which can ensure the concentration of micro-organisms responsible for bioconversion. Other methods include the pre-treatment of organic waste, the mixing of different waste types for digestion, the addition of chemicals, etc.

Studies have shown that the addition of nickel in the low concentration to poultry waste significantly increased biogas yield. Biogas production increased four hours after the addition of the nickel because nickel is naturally present in layer excreta but in a form that is unavailable for bacteria except when added artificially.

The use of mixed substrates to enhance biogas generation is another suitable method of improving the anaerobic digestion process. The substrates can be mixed in various ratios deriving from experiences over time to get the expected maximum yield. The efficiency of the digestion process can be increased substantially by improving the availability of the digestible material to microorganisms by pre-treating the waste feedstock. Pre-treatment methods include mechanical methods (e.g., milling to diminish the particle size), chemical methods (e.g., swelling in an alkaline solution), physical methods (e.g., ultraviolet radiation), thermal methods (e.g., heating of the feedstock), and a combination of the above methods. It was found that diminishing particle size increases the rate but does not affect the ultimate degree of biodegradation. Similarly, chemical treatment with acid or alkaline for 1 hour at pH 1 and 13 appeared to be the most adequate method because it increased the degree of biodegradation of the waste by about 30%. Thermal treatment of the waste materials for 1 hour at 100°C under atmospheric pressure improved the bio-digestibility by 8%.

Advantages have been reported for re-concentrated flushed swine waste (waste to which fresh waste has been added to increase its TS concentration) as a method of improving anaerobic digestion of wastes. Results showed that re-concentrated flushed swine waste exhibits excellent properties while digesting at thermophilic temperature. The use of 'aged' culture (waste slurry of many weeks) allows heavy loading (about 70 VS/L) of swine waste at a short retention time of 8 days with no adverse effect. This is because the 'aged' culture (substrate) is rich in micro-organisms which caused a higher rate of bioconversion of the culture into biogas, thereby necessitating a quicker rate of addition of fresh waste.

Stirring or mixing of the digester content (substrate in the digester) enhances gas production by improving the anaerobic digestion process. Stirring or mixing prevents the settlement of solids, thus avoiding the formation of surface scum and maintaining an even temperature gradient within the digester. Mixing ensures that the solid retention time equals the hydraulic retention time and enables the three phases of gas formation to take place throughout the digester. This actively discourages the accumulation of floating or settled solids. Agitation of the digester content ensures high rate of gas production by improving heat transfer, distribution of micro-organisms and maintaining good fluid consistency for reliable outflow.

Mixing, however, has some disadvantages. Mixing may not improve the anaerobic digestion process because non-mixing enables the accommodation of high solid loading of the digester and provides the separation of microbial phases within a continuous-flow digester type to substantially complete bioconversion of biodegradable feedstock components. In a non-mixed digester, the gas bubbling up through the digester content to the headspace of the digester provides a gentle agitation of the digester content without disrupting the solid concentration gradient of the digester. Another disadvantage of mixing the digester content during the anaerobic digestion process is that non-mixing ensures that contaminants introduced into the digester remains at localized points and are less likely to poison the entire digester volume, whereas contamination of the entire digester volume occurs rapidly in stirred digesters due to continuous mixing. Energy costs related to mechanical mixing equipment and its maintenance are eliminated by not mixing the digester content.

The two-phase concept of anaerobic digestion of wastes has also been found to improve the anaerobic digestion process and enhance greater gas yield when compared to conventional anaerobic bio-digestion methods. The two-phase concept consists of first digesting the waste in an acid phase. The gaseous and liquid products of this phase are separately conveyed into separate methane fermenters to enable biogas production.

5.3.1 Thermophilic Digestion Using a Solar House

The gas yield from batch-type anaerobic digesters and even continuous-flow digesters can be improved by operating them at elevated temperatures of about 50°C and above. This is because digestion at the elevated temperatures (thermophilic) is faster and higher for any waste type due to reduced detention time of the substrate in the digester. Most tropical countries like Nigeria are blessed with plentiful sunshine all the year round. Nigeria, for instance, receives about 490 W/m²/day. There is thus plenty of potential for the development of solar energy in thermophilic digestion. The other advantages of this are low maintenance cost and no pollutants.

A solar house to trap heat can be built and the digester was placed in it. Heat is transferred to the slurry by conduction through the digester wall. Stones or rock pebbles can be placed in the solar house to absorb and store heat, which is released during low-temperature situations, particularly at night to maintain a steady diurnal temperature. The solar house can also be built over a floating-drum plant where heat is transferred to the slurry through the floating drum of the plant.

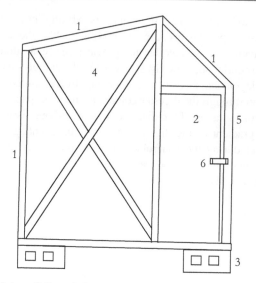

1. Transparent polythene; 2. Door; 3. Concrete block; 4. Black polythene; 5. Lumber; 6. Door staple

FIGURE 5.2   Cross-sectional view of the solar house.

Figure 5.2 shows the side view of a typical solar house. The solar house measures 2,700 mm in length, 1,830 mm in width and 2,300 mm in height. The house is built from lumber and covered with transparent and black 15 mil polythene films that are used as the walls and roof of the house. The films are double layers separated by 50 mm. The south-facing roof should be inclined between 10° and 15° to the horizontal for Nigeria locations to enable maximum collection of solar radiation incident on it. The roof and the other larger surface areas of the house should be covered with the transparent film, while the smaller surface areas (the sides) should be covered with the black film. The door, which is a double black cover, is positioned on the side of the house. The house should be oriented in such a way that the larger surface areas face the east and the west with the smaller surface areas facing the north and south. The house should be raised about 500 mm above the ground level to prevent water-related problems and improve its durability.

5.3.1.1 Quantity of Stones Required for Heat Storage in the Solar House

A computation of the quantity of stones required for storing heat within the solar house for use during low-temperature situations is detailed as follows:

Heat energy available to the house

The energy gained from the collection can be computed from equation 5.2

$$Q_n = AGC_p(T_i - T_n) \tag{5.2}$$

where

Q_n is the energy gained from the solar house, J.

G is the collector fluid mass flow rate per unit collector area.

C_p is the specific heat capacity of the collector fluid (air at 24°C).

T_i is the temperature in the house.

T_n is the temperature outside the house (ambient).

A is the collector area (m^2); the following constitute the collector area:

- The south facing roof that is tilted 15° to the horizontal.
- The entire enclosure with polythene film walling.
- The walls and roof have a double-layer collector.

It is assumed that the collector is 10% efficient. Therefore, the heat available to the solar house (HE) is 10% of Q_n.

Heat energy available for storage

It is assumed that 5% of the available heat to the solar house for 6 hours effective operation cycle is to be stored in stones placed inside the house. Therefore, the total energy available for storage (TEA) is 5% of HE.

Quantity of stones

The quantity of stones required for heat storage (M) can be computed from equation 5.3.

$$M = \frac{TEA}{C\Delta t} \tag{5.3}$$

where

C is the specific heat capacity of stone, J/kg°C

Δt is the observed difference between daily maximum and minimum temperature, °C

5.3.2 Use of Media Materials

The use of media materials is one of the methods of improving biogas yield from the anaerobic digestion of organic material such as poultry waste (Bolte et al., 1986; Hill and Bolte, 1986). Media materials are substances or surfaces on which anaerobes can be attached and concentrated such that the bioconversion of organic material into biogas is enhanced. Experiments were first carried out using media materials in anaerobic digestion in 1984 by Blanchard and Bolte. Materials such as stone, plastic, polyurethane and wood have been used as media material. Media material was first introduced into the anaerobic digestion process as a means of reducing bacteria 'washout' from continuous-flow digesters. The use of media material in a batch digestion process eliminates the need for stirring or mixing the digester content. This is because the well-distributed media materials containing a reasonable population of anaerobes enables all the phases of gas production to be undertaken at all points within the digester. The media materials also help to ensure rapid start-up of the digestion process after the digested slurry is poured out of the batch digester, thus eliminating the need for 'seeding' of the freshly loaded slurry.

Studies have demonstrated the effective use of wood as media materials in the biogas production process. Wood is a much better media material than commercial plastic because it is capable of shorter start-up periods and contrary to expectation showed no apparent decomposition after 52 weeks of operation. Wood contains a high amount of lignin which is not anaerobically degradable. Higher rates of substrate utilization were also indicated for wood media because of the lower volatile acid levels in the digester. A biofilm taken from *Obeche* (*Triplochiton scleroxylon*), an affordable Nigeria soft wood species used as media material, showed a high concentration of methanothix species of bacteria.

Figure 5.3 shows a typical batch-type anaerobic digester with media materials. The media materials are fixed and suspended within the digester.

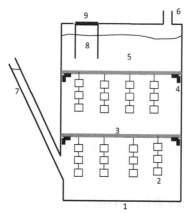

9	Influent pipe cover	1	Mild steel	Φ50 mm	
8	Influent pipe	1	Mild steel	Φ60 mm	
7	Effluent pipe	1	Mild steel	Φ300 mm x 1000 mm	
6	Gas pipe	1	Mild steel	Φ20 mm	
5	Slurry	-	Poultry waste	-	
4	Fixed media material holder	8	Mild steel	50 mm x 10 mm	Welded to the digester
3	Fixed media material	2	Obeche wood	Φ600 mm	Has Φ20 mm holes
2	Suspended media material	-	Obeche wood	25 mm x 15 mm x 5 mm	
1	Anaerobic digester	1	Mild steel	Φ600 mm	
Part No.	Component name	Qty.	Material	Dimension	Remark

FIGURE 5.3 Cross-sectional view of a batch anaerobic digester containing media materials.

These materials constituted 2% of the hydraulic volume of the digester. The fixed media material is 600 mm in diameter with holes of 20 mm to allow for the gas produced to bubble through the digester headspace. The two fixed media materials are placed about 30 mm apart, that is, at 350 and 650 mm from the bottom of the digester. The fixed media material is held in the position within the digester by means of holders which are welded to the digester. The suspended media materials are each $25 \times 15 \times 5$ mm with a hole of 3 mm. They are held in suspension within the digester by passing a thread through the 3 mm hole and tying this to the fixed media material.

The media fill ratio (MFR) of a digester is determined from equation 5.4

$$\text{MFR} = \frac{\text{Total volume of media materials}}{\text{Hydraulic volume of digester}} \times 100\% \qquad (5.4)$$

5.4 PURIFICATION OF BIOGAS

It is desirable to remove the carbon dioxide and hydrogen sulphide contents of biogas for sensitive uses, particularly for industrial applications. The gas can be purified by passing it through a lime solution or water to remove the carbon dioxide, thereby improving its heating value (Idikwu, 2013). Bubbling the gas through an aqueous solution of monoethanolamine reduces the carbon dioxide content of biogas by up to 0.5%–1.0% by volume. The hydrogen sulphide is removed by bubbling the gas through sawdust impregnated with iron filling in a mixture of 1:1 to prevent corrosion and mechanical wear of the equipment in which the biogas is used.

Purified biogas emits fewer amounts of greenhouse gases (Table 5.1) when used in internal combustion engines. Studies have shown that the carbon dioxide emission from use of un-purified, water-purified and calcium chloride-purified biogas is low and not significantly different because of the complete combustion of the fuels in the spark ignition engine. These studies showed that carbon dioxide emission from purified and un-purified biogas fuels were not significantly influenced by purification of the gas.

Several studies have also shown that the carbon monoxide emission is significantly lower than that from the un-purified biogas, which may be because of the higher carbon dioxide content of the un-purified biogas. Many studies have established that higher carbon dioxide content of biogas results in higher carbon monoxide emission when used in spark ignition engines because of the incomplete combustion of the biogas.

Water purification yields the lowest sulphur dioxide emission because it reduces the hydrogen sulphide content of the gas. The hydrogen sulphide content of the gas is better dissolved in water than in calcium chloride. Lower

TABLE 5.1 Greenhouse gas emissions from gasoline, un-purified and purified biogas

GHG EMISSION (PPM)	UNPURIFIED BIOGAS	WATER-PURIFIED BIOGAS	$CACL_2$-PURIFIED BIOGAS	GASOLINE
CO	407[a]	014[a]	043[a]	1052[b]
CO_2	2.8[a]	3.4[a]	2.8[a]	1.8[a]
NO_x	00.01[a]	00.07[b]	00.10[b]	00.10[b]
SO_2	0.27[b]	0.13[a]	0.33[b]	0.33[b]

Source: Itodo et al. (2019).
Means with the same letter along the same row are not significantly different at $p \leq 0.05$ using the Duncan's New Multiple Range Test (DNMRT).

concentration of hydrogen sulphide in biogas yields low sulphur dioxide emission when biogas is used as fuel. The solubility of hydrogen sulphide in water is 2.5 g/kg of water, which is the highest solubility of all gas constituents of biogas. The solubility of the other gas constituents of biogas is 0.02, 0.03, 0.02, 0.001, and 0.01 g/kg for methane, oxygen, carbon monoxide, hydrogen, and nitrogen, respectively. Water purification of biogas gives the least carbon monoxide and sulphur dioxide emissions.

A gas purifier can be constructed using two detachable 2 L transparent plastic water filters with carbon filter string and an inlet and outlet pvc pipe at the top. The carbon string filters can be removed to create the gas purification cups (Figure 5.4). A 4.5 cm diameter pvc pipe is used to connect

FIGURE 5.4 Set-up of the biogas purifier.

the outlet of the first gas purification cup to the inlet of the second purification cup. The pvc pipes to the inlet of the first purification cup and outlet of the second purification cup will be fastened with copper nipple adopter to enable the attachment of the gas hose from the biogas plant to the first purification cup and the outlet pipe from the second purification cup to the hybrid carburetor of the generator. The set-up of the gas purifier can be fastened to a wooden stand. Biogas from the biogas plant will flow through the solutions in the purification cups into the hybrid carburetor attached to the engine.

Biogas Plants

6

Biogas plants are recognized as one of the most decentralized sources of energy supply as they are cheaper than large power plants. The most widely built biogas plants across the world are the Indian type continuous-flow floating drum and the Chinese fixed dome digesters. The floating-drum plant is to the Indians what the fixed dome plant is to the Chinese.

6.1 THE FLOATING-DRUM PLANT

The floating-drum continuous-flow biogas plant consists of a digester well that contains the slurry and a moving gasholder that floats either directly on the fermentation slurry or a water jacket on top of the digesting slurry. The slurry is fed into the digester through an inlet pipe. The digested slurry exits the digester through an effluent pipe into an effluent pit. The slurry is mixed in an influent pit that opens into the influent pipe. The drum rises when gas is collected in it and falls when the gas is drawn off from it. The gas drum is prevented from tilting by a guide frame called the drum holder.

This type of plant is simple and requires lesser skill to operate and construct than the fixed dome plant and provides a constant gas pressure. It requires the application of weights on the drum in order to deliver gas at the required pressure of 700–2,000 pa, which is the recommended pressure for utilizing biogas in a burner. The disadvantage of this type of plant is its high construction cost and the corrosion of the floating drum resulting from its short life of less than 5 years in tropical coastal regions. Also, cost is incurred in the regular maintenance of the steel gasholder because of painting of the drum. Figure 6.1 shows the diagram of a floating-drum plant. The floating-drum plant is commonly referred to as the Indian floating-drum plant and is the most popular type of biogas plant in India.

DOI: 10.1201/9781003241959-6

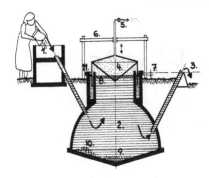

1. Influent mixing pit; 2. Digester well; 3. Overflow on effluent pipe; 4. Floating gasholder with braces; 5. Gas outlet pipe; 6. Gasholder guide; 7. Difference in level = gas pressure in cm WC8. Scum; 9. Sludge; 10. Grit and stones; 11. Water jacket.

FIGURE 6.1   A floating-drum biogas plant.

6.2 THE FIXED DOME PLANT

The fixed dome plant is made up of a digester with a fixed, non-moveable gas space at the upper part of the digester. When gas is produced it displaces slurry into the compensating tank. The gas pressure increases with the volume of gas stored. When gas production is low, there will be little gas in the digester head-space and the gas pressure will be low. The gas pressure becomes too high if the gas production is high. The advantage of this plant is that it has low construction cost and has a long life as it has no moving part. This type of digester is recommended only where construction can be supervised by experienced biogas technicians. Figure 6.2 shows the diagram of a fixed dome plant. The fixed dome plant is most often referred to as the Chinese fixed dome plant and constitutes the majority of biogas plants in China.

6.3 THE BALLOON PLANT

A balloon plant is a heat-sealed plastic or rubber bag (balloon) (Figure 6.3). Slurry is fed into the balloon and the gas produced is held at the headspace of the balloon from where the gas is piped out for use. The slurry is mixed in an influent pit from where it flows into the bottom portion of the balloon.

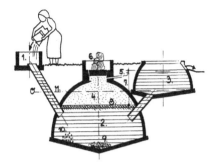

1. Digester well; 3. Compensating and removal tank; 4. Gasholder; 5. Gas pipe; 6. Entry hatch with gastight seal and weighted; 7. Difference in level = gas pressure in cm WC; 8. Supernatant scum; 9. Accumulation of thick sludge. 10. Accumulation of grit and stones; 11. Zero line: filling height without gas pressure.

FIGURE 6.2   A fixed dome biogas plant.

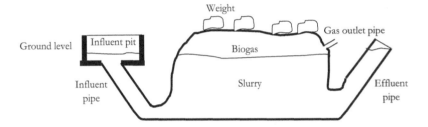

FIGURE 6.3   A balloon digester.

The digested slurry exits the balloon through an effluent pipe fitted to the bottom of the balloon. Gas pressure can be increased by the addition of weights on the balloon. Safety valves are installed to avoid excess pressure that will burst the balloon if its pressure limit is exceeded. Gas pumps may be used if higher pressures are required for a particular application. Reinforced plastic balloons are preferred because they are weather and ultraviolent ray resistant.

The advantages of this plant are standard pre-fabrication at low cost, shallow installation suitable for use in areas with high groundwater table, high digester temperatures in warm climates, uncomplicated cleaning, emptying and maintenance. The disadvantages of this plant are low gas pressure, which may require the use of pumps, scum cannot be removed during operation, the relatively short useful lifespan of the balloon because of its susceptibility to mechanical damage and it is usually not locally available. In addition, local craftsmen are rarely in a position to repair a damaged balloon. Balloon biogas plants are recommended if local repair is or can be made possible and the cost advantage is substantial.

6.4 STORAGE OF BIOGAS

Biogas can be stored at low, medium and high pressures. The low pressures (<0.005 MPa) are used for such applications as heating and in boilers. Biogas stored at the medium pressure (1–2 MPa) is used in stationary engines and irrigation pumps. The high pressure (>20 MPa) is used in stationary engines, irrigation pumps and as automotive fuel (Waihyo, 2015; Yusuf, 2010).

A typical example of the low-pressure system is the floating-drum gas-holder of the floating-drum plant. The advantage of the low-pressure system is that safety conditions of these tanks are less important. A high-pressure system is considered unfit for agricultural biogas plants because of the accessories required. Safety provisions must be made for storing in medium-pressure tanks. Compressors are used in medium- and high-pressure storage and the gas pressure is reduced back to about 0.005 MPa before use. Figure 6.4 is a picture of balloon storage for a fixed dome plant at the Nigerian Correctional Centre, Kirikiri, Lagos, Nigeria.

The choice of a blower or compressor will depend on the amount of pressure increase required by a system. Blowers are used to overcome piping pressure drop or for filling low-pressure storage tanks. Compressors are used to obtain medium and high pressures. Some medium-pressure compressors that handle small biogas flows are called boosters.

The energy required to compress biogas into a storage tank is a major operating cost of a biogas system. Therefore, estimating the energy requirement is an important component of designing a biogas system. The energy required to compress biogas into a storage tank can be estimated from equation 6.1.

$$W = C_1 R T_1 \left[\left(\frac{P_2}{P_1} \right)^{C_2} - 1 \right] \tag{6.1}$$

FIGURE 6.4   Balloon storage of a biogas plant at the Nigerian Correctional Centre, Kirikiri, Lagos, Nigeria.

where
W is shaft work required for compression, Pa
C_1 and C_2 are obtained from equations 6.3 and 6.4, respectively

$$C_1 = \frac{k}{(k-1)} \tag{6.2}$$

$$C_2 = \frac{(k-1)}{k} \tag{6.3}$$

k is the ratio of specific heats of biogas, C_p/C_v which is 1.3
    R is the gas constant for biogas
    T_1 is the initial temperature, °C
    P_1 is the initial pressure, Pa
    P_2 is the final pressure, Pa

Compressors are never 100% efficient because of friction and heat transfer that occur during the compression process. Therefore, the actual energy required is greater than the computed energy. Manufacturer's literature will indicate the efficiency of the compressor. Table 6.1 shows the different power requirements for the desired final pressure when compressing biogas.

TABLE 6.1   Power requirements for compressing biogas

PSIA	FORCE kPa	MPa	POWER HP	kW
19.8	136.52	0.14	0.72	0.54
50.0	344.74	0.34	0.98	0.73
75.0	517.11	0.52	1.17	0.87
100.0	689.48	0.69	1.33	0.99
125.0	861.0	0.86	1.46	1.09
175.0	1,206.58	1.21	1.57	1.17

Source: Heisler (1981).

Design of Biogas Plants

7

7.1 DESIGN ANALYSIS OF A FLOATING-DRUM PLANT

The size of the biogas plant, which is the volume of the digester well for a continuous-flow floating-drum plant, can be determined from equation 7.1.

$$V_d = RT \times LR \qquad (7.1)$$

where

V_d is the volume of the digester, m³

RT is the hydraulic retention time, day. A RT of 30–40 days is recommended for optimum biodegradation of the waste in the plant.

LR is the loading rate of waste into the plant, m³/day.

The LR is equal to the daily substrate input quantity into the plant (L_d) if the plant is loaded daily:

$$LR = L_d$$

If the plant is loaded once in every 7 days:

$$LR = L_d / 7$$

If the plant is loaded twice every 7 days:

$$LR = 2L_d / 7$$

DOI: 10.1201/9781003241959-7

47

The L_d is determined from equation 7.2.

$$L_d = B + W \tag{7.2}$$

where
  B is the daily volume of waste input into the plant, m^3
  W is the daily volume of water mixed with the waste, m^3

The ratio of waste to water (B:W) depends on the desired consistency of the slurry to be biodegraded in the plant. Most often, a B:W = 1:3 is used. The daily quantity of the waste input into the plant can be determined from equation 7.3.

$$B = \frac{G_u}{G_y \times \rho_w} \tag{7.3}$$

where
  G_u is the daily biogas consumption rate for the proposed application, m^3. The G_u for various applications are contained in Table 7.1.
  G_y is the biogas yield per kg of the waste type, m^3/kg. The biogas yield from various waste types is contained in Table 7.2.
  ρ_w is the density of the waste, kg/m^3. The density of $1,000\,kg/m^3$ can be used in designing plants that use cattle and piggery wastes, while the density of $450\,kg/m^3$ can be used to design plants that use poultry wastes.

The measurement for a number of biogas plants is contained in Table 7.3.

TABLE 7.1   Biogas consumption rate for various uses

USE	BIOGAS CONSUMPTION RATE (M³/MIN)
Boiling of water	0.7
Cooking of rice	2.8
Cooking of beans	4.9
Electricity generation	0.03

Source: Itodo et al. (2007a).

TABLE 7.2   Biogas yield from various waste types in Nigeria

WASTE TYPE	BIOGAS YIELD (M³/KG)
Cattle	0.03
Pig	0.05
Poultry	0.08

Source: Itodo & Awulu (1999).

TABLE 7.3 Measurement for a number of biogas plants

PLANT SIZE (M³)	NO. OF ANIMALS	WATER: WASTE PER DAY (1:1) (KG)	VOL. OF WELL FOR 40 DAYS DIGESTION (M³)	NO. OF BRICKS	NO. OF CEMENT BAGS (50 KG)	QTY. OF SAND (M³)	GAS PROD. PER DAY (M³)	QTY. OF FERTILIZER PER DAY (KG)	NO. OF PERSONS SERVED (COOKING+LIGHTING)
2	4	80	3.5	2,800	22	9	2	4–8	4–6
3	6	120	5	3,200	25	12	3	6–12	6–8
4	8	160	7	4,000	28	12	4	8–16	9–11
5	10	200	8.5	4,000	30	14	5	10–20	12–15
7.5	15	300	13	5,200	32	16	7.5	15–30	15–20
10	20	400	17	6,400	35	18	10	20–40	20–30

7.2 VOLUME OF THE GASHOLDER

Generally, 40%–60% of the daily gas production (G_y) has to be stored (Werner et al., 1984). The volume of gasholder (V_g) is determined from equation 7.4.

$$V_g = 0.6G_y \qquad\qquad (7.4)$$

The ratio of the volume of the gasholder to the volume of the digester (V_g:V_d) is somewhere between 1:3 and 1:10 with 1:5–1:6 occurring most frequently.

7.3 SCALING OF THE GASHOLDER

The gasholder capacity (C) is the ratio of the gasholder volume (V_g) to the daily gas yield from the plant (G_y). It is an important planning parameter in the design of floating-drum biogas plants. If the gasholder capacity is insufficient, part of the gas produced will be lost and the remaining volume of gas will not be enough. The cost of constructing a very large gasholder is unnecessarily high although operating the floating-drum plant is more convenient than the other type of plants.

Example, if a biogas plant is designed to be used 8 hours in a day, which represents about 33% of the estimated daily gas production time from the plant, the gasholder should be designed to hold the gas produced in the balance of the 16 hours of the day that is not consumed. Therefore, an appropriate gasholder capacity will be 67%.

The gasholder capacity is represented by equation 7.5.

$$C = \frac{V_g}{G_y} \qquad\qquad (7.5)$$

The gasholder capacity for a floating-drum biogas plant with a gasholder volume of 2.5 m³ and a daily gas yield of 3.0 m³ is 0.83 (83.3%). The gasholder capacity value ranges from 75% to 125% of the estimated gas production (Sasse, 1988).

7.4 DIMENSIONS OF THE FLOATING-DRUM PLANT

Figure 7.1 shows the diagram of a continuous-flow floating-drum biogas plant showing the dimensions. Appropriate scaling of the plant diameter and height is important.

7.4.1 Diameter of the Digester Well of the Plant

The ratio of the diameter of the digester well to its height should be 1:3 (equation 7.6).

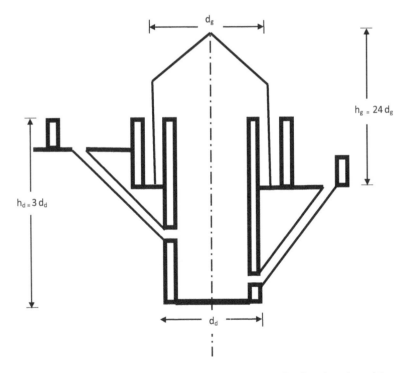

FIGURE 7.1   Appropriate sizing of the components of a floating-drum biogas plant.

$$h_d = 3d_d \tag{7.6}$$

where
  h_d is the height of the digester well of the plant, m
  d_d is the diameter of the digester of the plant, m

The volume of the cylindrical digester well of the plant is expressed in equation 7.7.

$$V_d = \frac{\pi \times d_d^2 \times h_d}{4} \tag{7.7}$$

Substituting equation 7.6 into 7.7, the volume of the digester well of the plant can be obtained from equation 7.8

$$V_d = \frac{3 \times \pi \times d_d^3}{4} = 2.36 d_d^3 \tag{7.8}$$

The diameter of the digester well of the plant can be determined from equation 7.9.

$$d_d = \sqrt[3]{\frac{V_d}{2.36}} \tag{7.9}$$

7.4.2 Height of the Digester Well of the Plant

The height of the digester well can be determined from equation 7.10 by substituting for d_d as in equation 7.6.

$$h_d = 3\sqrt[3]{\frac{V_d}{2.36}} \tag{7.10}$$

7.4.3 Diameter of the Gasholder

The ratio of the diameter of the gasholder to the height of the gasholder (h_g) for a floating-drum plant is expressed in equation 7.11.

$$h_g = 2.4 d_g \tag{7.11}$$

The volume of the gasholder can be determined from equation 7.12.

$$V_g = \frac{\pi d_g^2 h_g}{4} \tag{7.12}$$

Substituting for h_g (equation 7.11), the volume of the gasholder can be obtained from equation 7.13.

$$V_g = \frac{2.4 \times \pi \times d_g^2 \times d_g}{4} = 1.89 d_g^3 \qquad (7.13)$$

The diameter of the gasholder can be determined from equation 7.14.

$$d_g = \sqrt[3]{\frac{V_g}{1.89}} \qquad (7.14)$$

7.4.4 Height of the Gasholder

The height of the gasholder can be determined from equation 7.15.

$$h_g = 3\sqrt[3]{\frac{V_g}{1.89}} \qquad (7.15)$$

7.5 SIZE OF A FIXED-DOME PLANT

The digester capacity of a fixed-dome plant can be determined from equation 7.16.

$$V_{dd} = V_g + V_s \qquad (7.16)$$

where
V_{dd} is the volume of the fixed-dome plant, m^3
V_g is the maximum volume of the gas yield from the plant, m^3
V_s is the maximum volume of the slurry in the plant, m^3

7.6 DESIGN OF A FLOATING-DRUM BIOGAS PLANT (SAMPLE CALCULATION)

Task: Design a floating-drum biogas plant to power a 3.0 kVA generator to generate electricity to light a poultry house for 12 hours at night using waste from the poultry.

What quantity of biogas is required to power the generator for 12 hours?
1.8 m³ of biogas is required to power the generator in 1 hour (Table 7.1). Therefore, 21.6 m³ of biogas will be required to power the generator in 12 hours.

What quantity of poultry waste is required to produce 21.6 m³ of biogas?
1 kg of poultry waste yields 0.08 m³ of biogas (Table 7.2).

0.08 m³ of biogas is produced from 1 kg of waste
1 m³ of biogas will be produced from 1/0.08 kg of waste
21.6 m³ of biogas will be produced from 1/0.08×21.6 kg of waste = 270 kg of waste

The density of poultry waste is 450 kg/m³
270 kg of poultry waste = 0.60 m³ of poultry waste
What is the loading rate of the waste into the digester?
The ratio of waste-to-water (B:W) is 1 part of waste to 3 parts of water
B:W = 1:3 = 0.6 m³: 1.8 m³

The quantity of the substrate (waste mixed with water) input into the plant (L_d) is 0.60 m³ + 1.8 m³ = 2.4 m³

The plant is to be loaded once in 7 days.
Therefore, the loading rate (LR) is:

$$LR = L_d/7 = 2.4/7 = 0.34 \text{ m}^3$$

The plant will be loaded with 270 kg (0.60 m³) of waste mixed with 1.8 m³ of water once in 7 days.
What is the volume of the digester well?
The volume of the digester well (V_g) is calculated using equation 7.1.

$$V_d = RT \times LR = 30 \times 0.34 = 10.3 \text{ m}^3$$

RT of 30 is used in determining the volume of the digester.
Diameter of the digester well
The diameter of the digester well was calculated using equation 7.9

$$d_d = \sqrt[3]{\frac{V_d}{2.36}} = d_d = \sqrt[3]{\frac{10.3}{2.36}} = 1.64 \text{ m}$$

The height of the digester well was calculated using equation 7.10

$$h_d = 3\sqrt[3]{\frac{V_d}{2.36}} = h_d = 3\sqrt[3]{\frac{10.3}{2.36}} = 4.9 \text{ m}$$

Volume of gasholder

The volume of gasholder was calculated using equation 7.4.

$$V_g = 0.6G_y = 0.6 \times 21.6 = 12.96 \text{ m}^3$$

Height of gasholder
The height of the gasholder was calculated using equation 7.15.

$$h_g = \sqrt[3]{\frac{V_g}{1.89}} = \sqrt[3]{\frac{12.96}{1.89}} = 3\sqrt[3]{6.7} = 3 \times 1.9 = 5.7 \text{ m}$$

Diameter of gasholder
The diameter of the gasholder was calculated using equation 7.14.

$$d_g = \sqrt[3]{\frac{V_g}{1.89}} = \sqrt[3]{\frac{12.96}{1.89}} = \sqrt[3]{6.7} = 1.9 \text{ m}$$

Payback period
Tables 7.4 and 7.5 list the table of returns and the cash flow data for the designed biogas plant, respectively. The total investment is N729,500.00. The payback period occurred before the first year because the return in the first year was greater than the total cost of investment. The payback period is thus calculated:

$$t_{pb} = 9 + \frac{I_o - \text{Cumulative returns just before the break} - \text{even}}{\text{Returns at the cumulative returns after the breakeven}}$$

$$t_{pb} = 1 + \frac{729,500 - 966,000}{966,000} = 1 - 0.24 = 0.76 \text{ years} = 9 \text{ months}$$

Return on investment
The return on investment for the designed plant is calculated thus:

$$\text{ROI} = \frac{\text{NP}}{I_A} \times 100$$

TABLE 7.4 Financial returns for the designed biogas plant

YEAR	RETURN (N)	CUMULATIVE PAYBACK (N)
1	966,000	966,000
2	966,000	1,932,000
3	966,000	2,898,000

TABLE 7.5   Data sheet for economic analysis of the designed biogas plant

Project title: A biogas plant for electricity location: Makurdi, Nigeria Owner: Mr. Kelvin Obida
Type of plant: A floating-drum continuous-flow plant digester volume: 4 m³ Type of waste: poultry waste
Estimated service life: 15 years

ITEM	YEAR															
	0	1	2	3	4	5	6	7	8	9	10	11	12	13	14	15
1.0 Investment Costs																
1.1 Planning, design, supervision consultancy	150															
1.2 Bricks	67.5															
1.3 Cement	84															
1.4 Sand	30															
1.5 Water	8															
1.6 Flat metal sheet	60															
1.7 Hybrid carburetor & accessories	20															
1.8 Electric generator	200															
1.9 Digging of digester well	40															

(Continued)

TABLE 7.5 (Continued)   Data sheet for economic analysis of the designed biogas plant

Project title: A biogas plant for electricity location: Makurdi, Nigeria Owner: Mr. Kelvin Obida
Type of plant: A floating-drum continuous-flow plant digester volume: $4\,m^3$ Type of waste: poultry waste
Estimated service life: 15 years

ITEM	0	1	2	3	4	5	6	7	8	9	10	11	12	13	14	15
1.10 Labour	70															
1.0 Total investment cost	729.5															
2.0 Income																
2.1 Energy revenue[1]		120	120	120	120	120	120	120	120	120	120	120	120	120	120	120
2.2 Sale of digester effluent		900	900	900	900	900	900	900	900	900	900	900	900	900	900	900
2.0 Total income		1,020	1,020	1,020	1,020	1,020	1,020	1,020	1,020	1,020	1,020	1,020	1,020	1,020	1,020	1,020
3.0 Expenditure																
3.1 Repair & maintenance		30	30	30	30	30	30	30	30	30	30	30	30	30	30	30
3.2 Transport of wastes[2]		24	24	24	24	24	24	24	24	24	24	24	24	24	24	24

(Continued)

TABLE 7.5 (Continued)   Data sheet for economic analysis of the designed biogas plant

Project title: A biogas plant for electricity location: Makurdi, Nigeria Owner: Mr. Kelvin Obida
Type of plant: A floating-drum continuous-flow plant digester volume: $4\,m^3$ Type of waste: poultry waste
Estimated service life: 15years

ITEM	0	1	2	3	4	5	6	7	8	9	10	11	12	13	14	15
3.0 Total expenditure		54	54	54	54	54	54	54	54	54	54	54	54	54	54	54
4.0 Returns (2.0–3.0)		966	966	966	966	966	966	966	966	966	966	966	966	966	966	966
5.0 Depreciation		48.6	48.6	48.6	48.6	48.6	48.6	48.6	48.6	48.6	48.6	48.6	48.6	48.6	48.6	48.6
6.0 Net profit (4.0–5.0)		917.4	917.4	917.4	917.4	917.4	917.4	917.4	917.4	917.4	917.4	917.4	917.4	917.4	917.4	917.4

The amount is in thousands of the Nigeria currency Naira.
[a] The amount that should have been paid as electricity tariff for use of grid electricity that is now being saved by use of the plant.
[b] The cost of hiring a tractor per hour once in a month for transport of waste to the site of the plant.

TABLE 7.6   Specifications of the designed biogas plant

DESIGNED PARAMETER	SYMBOL	VALUE
Volume of biogas required to power the generator for 12 hours	G_u	21.6 m³
Quantity of waste required to produce the needed amount of biogas	B	270 kg (0.60 m³)
Waste: Water	B:W	1:3 = 06:1.8 m³
Quantity of slurry input into the plant	L_d	B + W = 2.4 m³
Waste loading scheme		Once in 7 days
Loading rate of slurry into the digester	LR	0.34 m³
Retention time	RT	30 days
Volume of digester well	V_d	10.3 m³
Diameter of the digester well	d_d	1.64 m³
Height of the digester well	h_d	4.9 m
Volume of gasholder	V_g	12.96 m³
Height of gasholder	h_g	5.7 m
Diameter of gasholder	d_g	1.9 m
Payback period	t_{pb}	9 months
Return on investment	ROI	251.5%

$$I_A = \frac{\text{Initial investment}}{2} = \frac{I_o}{2} = \frac{729,500}{2} = 364,750$$

$$\text{ROI} = \frac{\text{NP}}{I_A} \times 100 = \frac{917,366}{364,750} \times 100 = 251.5\%$$

The excellent payback period and very high ROI of the plant was because of the high income accruing from the sale of about 9,000 kg of its effluent at N5,000 per 50 kg bag as fertilizer. Table 7.6 lists the specifications of the designed biogas plant.

Construction of Biogas Plants

8

8.1 PRE-CONSTRUCTION CONSIDERATIONS

The followings should be considered before embarking on constructing a biogas plant:

8.1.1 Availability of Feedstock to Meet the Daily Need of Manure to Be Fed into the Digester

- The amount of manure fed daily into a digester is determined by the volume of the digester itself, divided over a period of 30–40 days.
- 30 days is chosen as a minimum time frame for optimum bacterial decomposition to take place to produce biogas and destroy many of the toxic pathogens found in wastes.

8.1.2 Location

- The digester pit should be dug at least 13 m away of a drinking water well.

DOI: 10.1201/9781003241959-8

- The inside of the digester well should be plastered with mortar if the water table is reached when digging the digester pit.
- The digester should be located close to the source of the waste to avoid the time and cost of transporting the waste.
- Determine whether there is enough space to build the plant.
- Be sure that water is readily available for mixing with the waste.
- Provision should be made for storing the slurry.
- The site for constructing the plant should be under a shade and exposed to the sun.
- The plant should also be as close as possible to the point of use of the gas.

8.1.3 Sizing of Plant

What Will Be the Volume of the Digester That Will Contain the Determined Amount of Slurry?

- Suppose a waste to water is 1:1.
- Six cattle will give 60 kg/manure+60 kg/water=120 kg.
- The total input per day will be 120 kg.
- The input for six weeks (42 days) will be 120 kg×42 days=5,040 kg.
- The rule of the thumb is 1,000 kg=1 m³.
- Therefore, 5,040 kg is 5.4 m³.
- The minimum capacity of the fermentation well is 5.4 m³.

What Is the Volume of the Digester Well?

- Assumes that the earth is not too hard to dig out and that the water table is low even in the rainy season.
- An approximate size for the 6 m³ tank would be a diameter of 1.5 m. Therefore, the depth required is 2.3 m.

What Is the Volume of the Gasholder?

- The drum is to hold between 60% and 70% of the total daily gas production.
- Therefore, the volume of the gas holder will be 70%×4 m³=2.8 m³.
- If the gas holder is of diameter 1.5 m, the height of the gas holder will be 1.1 m.

8.1.4 Material of Construction

The materials required for construction of the 4 m³ plant are:

- Baked bricks, approximately 4,000.
- Cement, foundation and wall covering, 28–40 bags of (50 kg) cement.
- Sand – 12 m³.
- Copper wire screen (25×25 cm).
- Rubber or plastic hose.
- Gas outlet pipe – 3 cm in diameter.
- Mild steel sheeting – 0.32 mm (30 gauge) to 1.63 mm (16 gauge) –9 m.

8.1.5 Tools

- Welding equipment for construction of the gasholder, pipe fittings, etc.
- Shovels for concrete and masonry works.
- Metal saw and blades for cutting steel.

8.2 CONSTRUCTION OF PLANT

Figure 8.1 shows pictures of the construction of a floating-drum biogas plant at the University Teaching and Research Farm, University of Agriculture, Makurdi, Nigeria.

8.2.1 Foundations and Walls

- Dig a pit 1.5 m in diameter to a depth of 2.5 m.
- Line the floor and walls of the pit with baked brick bound with lime mortar or clay.
- Make a ledge or cornice at two-thirds the height of the pit from the bottom. The ledge should be some 15 cm wide for the gas cap to rest on when empty.
- Extend the brickworks 30–40 cm above the ground, to bring the total depth of the pit to approximately 3 m.

FIGURE 8.1 Pictures of the construction of a floating-drum biogas plant in Makurdi, Nigeria.

- Build the ledge up to the height of the brickwork extension above the ground. This forms the space to be filled with water in which the gas holder with sit in and float.
- Put in place the 20 cm diameter influent and effluent clay pipes. Place the influent pipe 70 cm above the bottom of the pit.
- Place the effluent pipe 40 cm above the bottom of the pit opposite the influent pipe and end at the ground level.
- Put copper screening of 0.5 cm holes at the mouth of the influent and effluent pipes to prevent unwanted materials from entering the digester well.

8.2.2 Gasholder

- The gasholder should be constructed from 1.63 mm (16 gauge) mild steel sheets.
- The height of the drum should be 1/3 the depth of the pit.
- Make the diameter of the drum 10 cm less than that of the pit.
- Cut a 3 cm hole on the cap.
- Fix a rubber hose on the 3 cm hole.
- Paint the inside and outside of the drum with a coat of paint or tar.
- Ensure that the drum is air tight by filling with water to check for leakage.
- Handles should be welded to either side of the gasholder for lifting it out of the water jacket during repair.

8.2.3 Mixing and Effluent Pits

- Build the influent mixing pit near the outside opening of the influent pipe.
- Also, provide a pit at the outlet to catch the effluent.
- Make provision for drying the effluent when the plant is fully operational.

8.2.4 Commonly Encountered Problems in Constructing a Biogas Plant

The problems encountered in building a floating-drum biogas plant are presented:

1. Problem of digging of the digester well because of ponding of the dug well, even at a depth of 0.5 m. This required pumping of water during digging.
2. High water table that caused excessive seepage into the digester well of the plant even at a shallow depth of 1 m, necessitating 'thick' plastering of the digester well and monitoring the seepage rate for two raining seasons.
3. Incorrect construction of the water jacket, which resulted from the inappropriate dimension of the ledge on which the drum sits.
4. Inappropriate positioning of the influent and effluent pits of the plant.
5. The lack of technical depth of the technology made it challenging to make the plant functional.

8.3 CHECKING FOR GAS LEAK

Checking for gas leak is an important post-construction activity of a biogas plant because leaks lead to pressure loss at the point of use of the gas. Plant operators usually monitor biogas yield and quality; most don't check for gas leakage. Gas leakage causes huge financial loss to the use of biogas plants. The leakage hotspots are the valves, connection hoses, pipes, and welded joints on the floating-drum gasholder. It is best to conduct a detection survey before commencement of plant operations. Gas leakage of less than 1,001 CH_4/h is considered as minor while leakages of less than 10,001 CH_4/h gas leakages are categorized, as shown in Table 8.1. Leak-measurement computers/leak testers may be used to determine gas leakage from biogas plants. Methane-sensitive, laser and infra-red devices are also used to detect biogas leaks that are invisible to the naked eyes. In recent times, methane-sensitive cameras are used to detect leaks.

8.3.1 The Soapy Water Test

The soapy water test is a simple test used to detect biogas yield from biogas plants. The procedure is as follows:

1. Close the valve controlling gas from the gasholder to the appliance at the point of use.
2. Put some detergent into a bowl of water and mix properly.
3. Put the soapy water into a spray bottle or a sponge and apply to the gasholder drum, valves, pipes and hoses.
4. Open the valve for gas to flow through the system.
5. A leakage is indicated if bubbles are seen at any device or point or a rotten egg smell is perceived.
6. If a leak is not detected, rinse off the system with clean water and let it drip dry.

TABLE 8.1   Classification of gas leakage

CATEGORY	NO. OF POINTS OF LEAKAGE	MAINTENANCE
3	<3	Maintenance at the next regular revision of the plant
2	3–4	Within 6 months
1	5	To be fixed immediately

Operating a Biogas Plant

9

9.1 START-UP OF THE PLANT

1. $4\,m^3$ (4,000 L) of waste required to start-up a $6\,m^3$ new biogas digester.
2. About 20 kg of "seeder" should be added to get the bacteriological process started. The "seeder" can come from an existing biogas plant or sewage.
3. The waste should be mixed with an equal amount of water, and the seeder added, mixed, stirred and fed into the plant. Subsequently, 50 kg of water is added to 50 kg of waste, mixed and added to the plant daily.
4. It takes 4–6 weeks from when the time the digester is fully loaded for enough gas to be produced and the plant becomes fully operational.

9.2 REGULATING PLANT OPERATING PRESSURE

1. The pressure output from the floating-drum plant can be regulated by the addition and reduction of weight on the gas holder to increase the pressure of gas flow from the plant.
2. Increasing or decreasing the pressure affects the volume of gas delivered at the point of use.

9.3 MAINTENANCE AND REPAIR OF PLANT

Common problems encountered when operating biogas plants and the attendant maintenance practices are provided in Table 9.1.

TABLE 9.1   Common problems encountered and attendant maintenance practices

S/NO.	PROBLEM	POSSIBLE REASON(S)	SOLUTION
1	Gas drum will not rise	Scum formation No gas formed Leakage in the system	Patience. The system needs about 4–6 weeks to get properly started Stir the digester Dilute the digester by adding some water Ensure that there is no leakage in the system
2	No gas at the appliance	No gas formed Not enough pressure in the system to force the gas from the digester to the appliance Gas leakage	Consider solution #1 Adjust the inlet jet of the appliance
3	No gas formed	Toxicity in digester Inappropriate waste-water mixture	Flush out the content of the digester with water Add more waste to the digester
4	Inadequate quantity of gas being formed	Inappropriate waste-water ratio Slurry too thick or too thin Few populations of required microorganisms	Add a 'seeder' from another plant or a sewage to the digester Add more waste to the digester
5	Flame dies off too quickly at the appliance	Pressure from the digester too high	Adjust the inlet jet of the appliance Reduce the weight on the gas holder

9.4 SAFETY AND HEALTH CONSIDERATIONS

The constituent gases of biogas, methane, carbon dioxide, hydrogen, nitrogen, ammonia and hydrogen sulphide each contribute to health issues. Caution is required working with biogas. The following precautionary measures should be taken when working with biogas:

1. Adequate ventilation
2. Appropriate precautions
3. Good work practices
4. Engineering controls
5. Adequate personal protective equipment

Methane is not toxic but displaces oxygen in a confined space. It will collect at the upper space because it is lighter than air, thus causing an oxygen-deficient environment which can kill. High levels of carbon dioxide increase heart rate and respiratory rate, displaces oxygen supply in the bloodstream that can cause unconsciousness and eventually death. Hydrogen sulphide is a highly toxic gas. It causes eye irritation at low levels and smells like rotten egg. At high levels, it destroys the sense of smell and causes respiratory paralysis. At high levels, it is odourless and therefore does not show any warning. Table 9.2 shows the health hazards of the constituent gasses of biogas.

The hazards associated with the production and use of biogas include fire/explosion, asphyxiation and disease. Table 9.3 shows the hazards and safety issues of biogas plants. It was reported that about 800 biogas plant accidents occurred between 2005 and 2015 although only less than a dozen of these accidents resulted in human death. Hydrogen sulphide leakage from the charging hall of the biogas plant at Rhadereistedt, Germany resulted in the death of the truck driver and three other workers; a worker was injured and hospitalized.

1. Biogas with a methane content of approximately 60% forms an explosive mixture with 10%–30% air. Such an explosion occurred on a Canadian swine farm in 2003 (Choiniere, 2004). Hydrogen sulphide, ammonia and hydrogen also have the potential to explode. Consequently, no open flame should be used near a digester. Also, smoking near and around biogas plants is prohibited.

TABLE 9.2   Health hazards of the constituent gases of biogas

CONSTITUENT GAS	PROPERTIES	HEALTH HAZARD
Methane	Not toxic Lighter than air Causes oxygen deficiency	Suffocation
Carbon dioxide	Heavier than air Displaces oxygen supply in the bloodstream	Increases heart rate Increases respiratory rate Unconsciousness
Hydrogen sulphide	Highly toxic gas Smells like rotten egg at low level Odourless at high level	Eye irritation Destroys the sense of smell Causes respiratory paralysis Nausea, tears, headache, sleep loss, irritation of the lungs
Ammonia	Lighter than air Pungent odour	Irritates the eyes and respiratory tract Displaces oxygen in the bloodstream

TABLE 9.3   Hazards and safety considerations of a biogas plant

HAZARD	CAUSATIVE AGENT	SAFETY MEASURES
Fire/explosion	Methane, hydrogen sulphide, ammonia, Hydrogen	• No open flame should be used near the plant. • Smoking near and around the plant is prohibited. • The first drum full of biogas should not be ignited.
Asphyxiation	Biogas	• Do not enter the manure storage facility or where there is leakage of biogas without the adequate protective equipment.
Disease	Bacteria, fungi, viruses, parasites	• Personal protective equipment should be used in handling of the waste. • Wash hands before eating and drinking. • Avoid touching eyes, nose and other mucous membranes. • Keep the plant clean.

2. The first drum full of the gas may contain so much carbon dioxide that it will not burn. On the other hand, it may contain methane and air in the right proportion to explode. Therefore, the gasholder should be emptied and refilled again. The first drumful of gas should not be ignited.

3. Asphyxiation from biogas usually occurs in enclosed spaces where manure is stored. Three persons were reported to have died from asphyxiation from swine manure in an enclosure (Osbern and Crapo, 1981). Biogas practitioners are advised not to enter facilities for storing manure or where there is leakage of biogas.

4. Biogas is produced by the action of anaerobic bacteria on manure. The wastes used as feedstocks in the production of biogas contain bacteria, fungi, viruses and parasites that are infectious to humans. Protective personal equipment should be used to avoid contact with these wastes, particularly when loading into the plant. Wash your hands with soap and clean water before eating and drinking and should avoid touching the nose, eye and other mucous membranes. The plant should also be kept clean.

9.5 SAFETY AND HEALTH AUDIT

The planning, design and construction of a biogas plant is mostly based on technical and economic factors. However, health and safety factors are considered in the construction and operation of the plants. It is important to access the health and safety of an operating plant to the plant workers, visitors to the plant and users of the plant by undertaking a safety audit. A checklist of the health and safety audit of biogas plants are contained in Table 9.4.

TABLE 9.4 Checklist of the health and safety audit of biogas plants

S/NO.	ITEM	YES	NO	NA
1	Gas sensor meters are installed and working?			
2	Gas leakage detectors are installed and working?			
3	First aid box available and it is well stocked?			
4	Hand washing points in the vicinity of the plant?			

(Continued)

TABLE 9.4 (*Continued*)   Checklist of the health and safety audit of biogas plants

S/NO.	ITEM	YES	NO	NA
5	Appropriate personal protective equipment are available at appropriate places?			
6	Personal protective equipment available to visitor?			
7	Are workers using personal protective equipment? (Helmets, face masks, goggles, safety gloves, respirators, glasses)			
8	Is clean water available?			
9	Are no smoking signs visibly displayed?			
10	Are fire hazard signs visibly displayed?			
11	Are caution signs visibly displayed?			
12	Are there safety instructions at the appropriate places?			
13	Are route signs in place?			
14	Are fire extinguishers in appropriately placed and in good working conditions?			
15	Are sand buckets in place?			
16	Are there waste handling equipment?			
17	Are there slurry storage facilities?			
18	Are there manure storage facilities?			
19	Are the workers trained on safety of the workplace?			
20	Emergency service numbers posted?			
21	Are there biogas operational manuals?			
22	Are there records of accidents at the plant?			

NA – Not applicable.

Biogas Stove 10

The use of biogas in cooking and heating begins with an efficient stove. Biogas stoves are relatively simple appliances which can be manufactured by local blacksmiths or metal workers. Stoves may be constructed from mild steel or clay. Clay burners are widely used in China and their performances have been satisfactory. However, a cooker is more than just a burner. It must satisfy certain aesthetic and utility requirements, which vary widely from region to region. Thus, there is no such thing as an all-round burner.

All gas burners follow the same principle. The gas arrives with a certain speed at the stove, a speed created by the given pressure from the gas plant in the pipe of a certain diameter. The jet at the inlet of the burner increases this speed, thus producing a draft which sucks air (primary air) into the pipe. The primary air must be completely mixed with the biogas by widening the pipe to a minimum diameter, which is in constant relation to the diameter of the jet. The widening of the pipe again reduces the speed of the gas. This diffuse gas goes into the burner head. The burner head is formed in such a way as to allow equal gas pressure everywhere before the gas/air mixture leaves the burner through the orifices at a speed only slightly above the specific flame speed of biogas. More oxygen (secondary air) is supplied by the surrounding air to enable final combustion.

If combustion is perfect, the flame is dark blue and almost invisible in daylight. Stoves are normally designed to work with 75% primary air. If too little air is available, the gas does not burn fully and part of the gas escapes unused. If too much air is supplied, the flame cools off, thus prolonging the working time and increasing the gas demand.

The main factors that affect the use of biogas as a combustible gas are the gas and air mixing rate, speed of the flame, ignition temperature and gas pressure. Biogas needs less air per cubic meter compared to liquefied petroleum gas. This means that for the same amount of air, more biogas is required. This therefore requires that gas jets for biogas should be larger in diameter. 1 L of biogas requires about 5.7 L of air for complete combustion while 1 L of butane and propane require 30.9 and 23.8 L of air, respectively.

DOI: 10.1201/9781003241959-10

The 2-flame burners are the most popular type. There are several types of biogas stoves in use across the world. An example is the Peking stove that is widely used in China and the Jackwal stove widely used in Brazil. The Patel Ge 32 and Patel Ge 8 stoves are widely used in India, and the KIE burner is used in Kenya. The efficiency of using biogas is 55% in stoves, 24% in engines and 3% in lamps.

10.1 DESCRIPTION OF A BIOGAS STOVE

Figure 10.1 shows a schematic diagram of a burner, while Figure 10.2 shows a picture of a burner. The main components of the burner are the injector, the air/gas mixing chamber and the burner. The injector is held in position by a nut and washer that is welded to the frame of the stove. The injector tapers into a nozzle of about $0.01\,mm^2$, which enters into the air/gas mixing chamber. The air/gas mixing chamber opens into the burner head. The burner head has 35 jets each of $0.03\,cm^2$ from which the gas can be ignited. The air/gas chamber is held in position by two brackets welded to the frame. The combustion of biogas is regulated by moving the injector into and out of the air/gas chamber, which regulates the amount of air that enters into the chamber. If the injector is moved deeper into the air/gas mixing chamber, the drift of oxygen into the burner is reduced, thus reducing combustion. On the contrary, when the injector is moved out of the air/gas mixing chamber, more oxygen enters into the burner, thereby increasing combustion. Table 10.1 contains the specifications of the burner. The frames and the stands are made from angle bars. A wall made from a metal sheet welded round the frame serves as a wind breaker. The stove is connected to the gas holding unit of the biogas plant by a rubber hose which carries biogas from the gas holder of the plant to the stove.

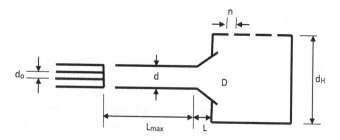

FIGURE 10.1   Schematic diagram of a burner.

FIGURE 10.2   Picture of a biogas stove.

TABLE 10.1   Specification of the biogas burner

COMPONENT	SYMBOL	SPECIFICATION
Jet diameter	D_o	1.6
Diameter of mixing pipe	D	9.7
Length of air intake hole		
Maximum	L_{max}	67.9
Minimum	L_{min}	13.1
Diameter of mixing chamber	D	12.6
Length of mixing chamber	L	14.6
Number of holes	N	35
Diameter of flame port hole	d_H	0.25

Source: Itodo et al. (2007).
All dimensions are in cm.

10.2 DESIGN OF A BIOGAS STOVE

The design of a biogas stove involves the determination of the following important dimensions:

1. Diameter of the jet (d_o)
2. Length of the air intake holes measured from the end of the jet (L_{max})

3. Length of the mixing pipe (L)
4. Number and diameter of flame port holes (d_H)
5. Height of the burner head (H)

10.2.1 Jet Diameter

The jet diameter (d_o) was estimated from equation 10.1

$$d_o = 2.1\sqrt{\frac{V_f}{\sqrt{h}}} \ (\text{mm}) \tag{10.1}$$

where
   V_f is the fuel flow rate (m³/h) and obtained from equation 10.2

$$Q = V_f = c\sqrt{\frac{\Delta p d^5}{SL}} \tag{10.2}$$

Q is the quantity of gas flow (m³/h)
   c is the pressure drop in the pipe, which for smooth plastic pipe is 2.80
   d is the diameter of the rubber hose, cm
   Δp is measured from the manometer (Figure 10.3), which is the allowed pressure drop, cm
   S is the air density = 1.2 kg/m³
   L is the distance between the manometers that is placed between the biogas plant and the stove to enable the measurement of pressure at which biogas

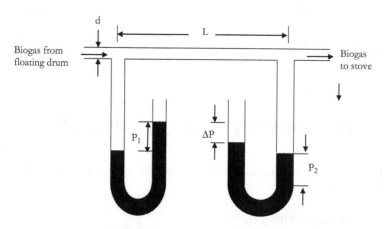

FIGURE 10.3   Set-up of the manometer for measuring pressure of gas entering the stove.

entered into the stove. The manometer is employed only if there is no gadget to measure the pressure.

h is the prescribed gas pressure (cm WC) which is 0.9 cm WC was determined from the manometer to be 0.01

10.2.2 Diameter of the Mixing Pipe

The diameter of the mixing pipe (d) is determined from equation 10.3

$$d = 6d_o \ (\text{mm}) \tag{10.3}$$

10.2.3 Length of Air Intake Holes

The maximum and minimum length of the air intake hole measured from the end of the jet is estimated from equations 10.4 and 10.5, respectively.

$$L_{\text{max}} \ (\text{mm}) = 7d \tag{10.4}$$

$$L_{\text{min}} \ (\text{mm}) = 1.35d \tag{10.5}$$

for a floating-drum biogas plant that is required to operate at a gas pressure of 0.10 m WG, the other dimension of the burner are determined as follows:

10.2.4 Diameter of the Mixing Chamber

The diameter of the mixing chamber (D) is determined from equation 10.6

$$D \ (\text{mm}) = 1.30d \tag{10.6}$$

10.2.5 Length of the Mixing Chamber

The length of the mixing chamber (L) is determined from equation 10.7

$$L = 1.50d \tag{10.7}$$

10.2.6 Number of Holes

The number of holes on the burner (n) is estimated from equation 10.8

$$n = 20d_o^2 \tag{10.8}$$

10.2.7 Diameter of the Flame Port Hole

The diameter of the flame port hole (d_H) for a floating-drum biogas plant is 2.5 mm

The cooking rate and efficiency of the stove are estimated from equations 10.9 and 10.10, respectively.

$$C_r = \frac{\text{Qty of commodity } (\text{g or L})}{\text{Time taken } (\text{min})} \tag{10.9}$$

$$\eta = \frac{C_r}{Q} = \frac{\text{Equation 11.9}}{\text{Equation 11.2}} \times 100\% \tag{10.10}$$

Utilization of Biogas and Effluent from Biogas Plants

11

Livestock wastes can be used to generate biogas to produce electricity and heat for heating in agricultural processing and other endeavours.

11.1 HEATING AND COOKING

Biogas is used mainly for heating and cooking purposes. It is a good replacement for fuel wood in many parts of the world among the rural poor for cooking. Biogas can be used in many heating applications such as:

a. Rearing of birds: Biogas generated from poultry wastes for heating in poultry pens.
b. Restaurants: Biogas plants fed with restaurant wastes to provide cooking gas (Figure 11.1).
c. Garri processing: Biogas plant fed with cassava peels to provide heating gas for garri-frying.

Experiences in boiling water with biogas in Nigeria show that $1\,m^3$ of biogas is used to boil $0.2\,L$ of water, which is far less than the $20\,L$ of water boiled with $1\,m^3$ of biogas in India. Similarly, only $0.001\,kg$ of rice is cooked with $1\,m^3$ of biogas in Nigeria, which is comparatively much lower than the $3.33\,kg$ of rice that is cooked with $1\,m^3$ of biogas in India (Table 11.1). The very poor performance of using biogas for boiling and cooking in Nigeria

DOI: 10.1201/9781003241959-11

FIGURE 11.1   Pictures of a floating-drum biogas plant in Makurdi, Nigeria and the burning flame from the plant.

TABLE 11.1   Amount of water and rice cooked per volume of biogas used in Nigeria compared to India

ACTIVITY	COUNTRY	COOKING RATE	BIOGAS USAGE RATE (M³/MIN)	AMOUNT COOKED PER M³ OF BIOGAS USED
Boiling of water	Nigeria[a]	0.14 L	0.69	0.2 L/m³
	India[b]	0.10 L	0.005	20 L/m³
Cooking of rice	Nigeria	5.13 g/min	4.78	0.001 kg/m³
	India	16.67 g/min	0.005	3.33 kg/m³

[a] Itodo et al. (2007).
[b] Werner et al. (1989).

may be because of poorly designed biogas stoves with efficiencies ranging between 20% and 53%. Furthermore, the biogas technology is still emerging in Nigeria with improving expertise and local capacity in the technology.

11.2 ELECTRICITY

Biogas can also be used in co-generation plants or purified to natural gas quality and fed into existing gas pipe lines for use in gas turbines to generate electricity. For example, electricity generation from biogas in Germany, US and UK is 15, 9 and 6 TWh, respectively.

The electricity demand in Nigeria is about 35,000 MW. Nigeria currently generates a theoretical 5,000 MW of electricity. There is thus a huge deficit that cannot be met from the existing power sources. The GHG emissions and high cost of gasoline and diesel fuels used in electric generators necessitate the use of cheap renewable fuels like biogas.

The use of biogas for electricity generation is very low in developing countries. Conversely, the main purpose for producing biogas in developed and industrialized countries is for electricity generation. The heat of combustion of biogas, which is a measure of its energy content, is 22 MJ/m^3 (15.6 MJ/kg), which means that 1 m^3 of biogas corresponds to 0.55 L of diesel fuel or about 6 kWh. The low calorific value of biogas is one of the barriers of biogas development in the combined heat and power generation.

The current practical option is the conversion of biogas to electricity by a generator. Biogas can therefore be used as fuel in combustion engines, which converts it to mechanical shaft power to produce electricity. Biogas can be used as fuel in the spark ignition (gasoline), compression ignition (diesel) engines, gas turbines and external combustion (Stirling) engines. Spark ignition engines can be operated on biogas alone. An advantage for the use of biogas in engines in contrast to natural gas is that it has high knock resistance property and can be used in engines with high compression ratios. However, biogas in spark ignition engines has smaller volumetric efficiency and therefore lowers power output. Biogas can be used in small generators (0.5–10 kW) and large power plants. The flameless combustion of biogas in combustion engines also has the advantage of combustion stability and low pollutant formation.

There are biogas engines, a type of spark ignition internal combustion engines that run on biogas or natural gas that operate like the Otto cycle. These biogas engines are, however, not commonly available and are expensive in developing countries like Nigeria. Consequently, biogas is being used in adapted engines of electrical generators to produce electricity. The engines of electrical generators are adapted to be fueled with biogas by use of the hybrid carburetor, which modifies the feeding, ignition systems and compression ratio.

In Germany, there are over 4,000 biogas plants running internal combustion engines to produce electricity. 439 TWh of electricity was generated from biomass sources in 2012. In Africa, the electricity generation from biogas is still limited to a few pilot plants. Kenya and South Africa are at the forefront of using biogas for electricity. The supporting factors for the use of biogas for electricity generation include the vulnerability of hydropower to drought, lack of awareness, capacity and experience. The use of biogas to produce electricity can only be considered economically viable if it competes favourably with electricity generation from fossil fuels and other renewables like hydropower.

Wastes from a livestock house used as feedstock in an attached biogas plant can be used to provide electricity for lighting the same house. It is estimated that 31.2 L/min of biogas is required to power a 3.2 kVA electric generator, which can be used to light livestock accommodation, power prime movers to produce mechanical shaft power for several other activities.

11.3 AUTOMOTIVE FUEL IN INTERNAL COMBUSTION ENGINES

Biogas as vehicle fuel can reduce GHG emissions by 60%–80% in the transport sector compared to fossil-based fuels like gasoline and diesel. When sufficiently purified, biogas can be used in place of fossil-based gas to drive natural gas vehicles (NGVs) or dual-fuel vehicles. Germany, Sweden, Switzerland, USA and UK were the largest producers of biogas vehicle fuel in 2016. Worldwide, about 500 plants produce 50 PJ per year of the upgraded biogas, often called bio-methane. Figure 11.2 shows the picture of a biogas-powered tractor exhibited at the International Biomass Conference Expo 2016 at the Convention Centre, Charlotte, North Carolina, USA.

The quality of biogas as fuel in spark-ignition engines is influenced by its calorific value, which is a function of its methane content, temperature and absolute pressure. Impurities in biogas, particularly water vapour, carbon dioxide and hydrogen sulphide, lower its calorific value, flame velocity, and flammability. The carbon dioxide in biogas reduces the nitrogen oxide and increases the carbon monoxide emissions. The increase in methane in biogas used as fuel in spark ignition engines significantly enhances the performance and reduces the emission of hydrocarbons.

The power output from unpurified biogas used as fuel in internal combustion engines is higher than that from purified biogas. The power output from unpurified biogas fuel compares favourably with that of gasoline in spark ignition engines, which may be because of a positive air/fuel ratio of the engine that allows the engine to run at its correct stoichiometric

FIGURE 11.2   Biogas-powered tractors exhibited at the International Biomass Conference & Expo, Charlotte Convention Centre, North Carolina, USA in 2016.

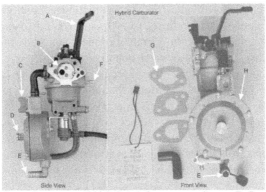

A. Choke B. Gasoline carburetor manifold C. Switch for gas or gasoline use D. Pressure release valve E. Gas supply inlet F. Gasoline supply inlet G. Gas and air mixing chamber

FIGURE 11.3 Picture of the hybrid carburetor and its parts.

ratio allowing complete combustion of the air-fuel mixture. The lower power output from purified biogas fuel may be as a result of the reduced volume of the fuel because of the removal of some of its constituents such as carbon dioxide that constitutes about 40% of its volume, which affects the air/fuel ratio of the fuel in the engine. This also explains the lower fuel consumption of purified biogas fuels in engines.

Biogas is used as fuel in spark ignition internal combustion engines by passing the gas through a hybrid dual-fuel carburetor (Figure 11.3) attached to the intake manifold of the spark ignition engine. The hybrid carburetor allows instantaneous switching to gasoline fuel without any other physical modifications. The hybrid carburetor ensures a smooth supply of biogas at low pressure to the engine based on the principles that when the engine is running, negative pressure enters into the air/fuel mixer via the pressure-regulating valve. The sealing arms of the swing valve open to supply the gaseous fuel after decomposition of the initial charge, so that the engine operates smoothly.

11.4 UTILIZATION OF EFFLUENT SLURRY FROM BIOGAS PLANTS

The effluent from the anaerobic digestion process of biogas production is a high-quality fertilizer because it is in the form of ammonium (NH^+_4), which readily attaches to the negative charge clay of the soil. It can effectively be used in the organic fertilization of peasant small farm holding that about in Nigeria. The sludge that is produced from the anaerobic digestion process is a better fertilizer and soil conditioner than either composted or fresh manure. This is because the liquid effluent contains many elements essential to plant life. It contains nitrogen, phosphorous, potassium and small amounts of metallic salts that are indispensable for plant growth. When the sludge is applied on the soil as fertilizer, its nitrogen is converted to ammonium ions (NH_4^+), which fix themselves to the negative charged clay particles of the soil, thereby making nitrogen available to the plants.

The effluent from the biogas plant can also be treated by heating to kill the microorganisms, mixed with grains and molasses and used as animal feed.

Renewable Energy Policies and Standards in Nigeria

12

The poor performance of installed renewable energy technologies (RETs) has militated against the acceptance and dissemination of the technologies in Nigeria. There is need for a department of standards in the Energy Commission of Nigeria (ECN), which should also be responsible for certification of RE artisans and practitioners. ECN should develop a template for RE standards in Nigeria. Regular and timely joint meetings between the Standards Organization of Nigeria (SON) and ECN are required to formalize the process for RE standards in Nigeria in line with their specific mandates. The successful dissemination of RE in Nigeria is impossible without first providing the standards because the standards are used to determine the correctness of the design and installation of RETs.

12.1 ENERGY POLICIES AND STANDARDS IN NIGERIA

Energy policies in Nigeria are championed by the ECN, learned societies for renewable energy in Nigeria and non-governmental organizations such as the Nigerian Energy Support Programme of the German International Corporation (GIZ).

12.1.1 The National Energy Policy

The National Energy Policy (NEP) was crafted by the ECN in 2003. The policy contains objectives and strategies for oil, natural gas, tar sand, coal, nuclear, hydropower, solar, biomass, wind and hydrogen. It has nothing on bio-fuels: bio-ethanol, biodiesel and biogas. The policies of fuelwood (page 16) and biomass (page 19) inadequately provided for the other biomass energy sources such as bio-fuels. A policy for each of these energy types creates a better environment for their promotion, exploitation and utilization taking into account their distinctiveness.

12.1.2 The Bio-Fuels Energy Policy and Incentives

The Bio-fuels Energy Policy and Incentives contains the policy objectives and strategies of bio-fuels in Nigeria although not in the format of the NEP. This policy is a Nigerian Gazette Number 72, Volume 94 of June 2007. The implementation of this policy is domiciled with the Department of Petroleum Resources (DPR). Bio-fuels are generally and erroneously seen as competitive substitutes to fossil gasoline and diesel, which are regulated by the DPR in Nigeria. Therefore, can the DPR truly champion the deployment of bio-fuels in Nigeria? Again, biogas is not captured in this policy! This document can be used to craft a policy for bio-fuels in the format of the NEP to eliminate the contradictions in the two important policy documents of the Federal Republic of Nigeria.

12.1.3 The Rural Electrification Strategy and Implementation Plan and Mini-Grid Electrification

This is an enabling framework to private investments by the National Energy Support Programme (NESP) of the GIZ. This framework contains the Nigerian standard for solar PV panels, batteries and charge controllers developed by the NESP of the GIZ. The Building Energy Efficiency Code contained in this policy is being gradually introduced in Nigerian states by the NESP of the GIZ.

12.2 RENEWABLE ENERGY STANDARDS IN NIGERIA

The pertinent issues in a standard are consensus, approval, recognized body, guidelines, procedures, performance criteria and goals.

A standard is a document established by consensus and approved by a recognized body that provides for common and repetitive use, the specifications, rules and guidelines for characterizing materials, processes, products, tests, testing procedures and performance criteria and their results in an effort to achieve certain specified goals at optimum degree of order in a given context. Standards can be mandatory or voluntary.

There are no biogas standards in Nigeria, which has made quality assurance of biogas technologies impossible. There is conflict in the development process and adoption of RE standards in Nigeria resulting from a misunderstanding of who is in charge, who has the right to develop standards and how many standards for a RE resource. Worldwide experience has shown that RE standards can emanate from professional associations, NGOs, relevant government agencies such as the ECN.

12.2.1 Objectives of Standardization

The objectives of standardization are to:

1. Simplify and reduce the growing variety and ever-increasing complexity of manufactured products and procedure in human life,
2. Provide a means of communication between the manufacturer and the customer, to list the items that are available, their size and performance,
3. Achieve an overall economy in labour, in the design and manufacture of goods,
4. Safety, health and protection of life,
5. Protect the interest of the consumer and the community and
6. Eliminate barriers in trade and tariff among nations.

12.2.2 Characteristics of Standards

1. Must be clear and unambiguous
2. Should be complete and as much as possible self-contained and self-explained

3. Should be implementable
4. Distinction should be made between voluntary and mandatory standards
5. Only a department or authority should be responsible for the management of standard
6. Standards are subject to periodic and orderly review
7. Distinction should be made between functional and non-functional characteristics
8. Should solve problems instead of creating them
9. Should be accessible and testable
10. Should promote interchangeability and uniformity

12.2.3 Developing Nigerian Biogas Standards

A Nigerian biogas standard must take into account raw materials, market requirements, availability of precise measuring instruments, availability of technology and local conditions.

12.2.4 Contents of the Standard

a. Forward
  • Who developed the standard?
  • What does the standard deal with? Expectation of the standard. E.g., provide guidelines and requirements to builders and users of biogas plant.
  • References used in formulating the standard.
b. Scope
  • This standard covers the construction, operation, maintenance and safety of biogas plants.
  • Terminology
    – Specify the definitions that apply in the standard
c. Body
  • Size of components (optimum dimensions).
  • Functional requirements.
  • Related issues.
  • Required proposal information, which is referred to as Form 1: Template for Renewable Energy Standards. This template includes relevant tables and figures, sample design, operation and maintenance, troubleshooting, environmental impact, etc.

Form 1-Template for RE standards: Required proposal information

1. Notes.
2. The issue: A concise statement describing the business, what the proposal seeks to address.
3. The scope of the issue.
4. Technological benefits.
5. Economic benefits.
6. Societal benefit.
7. Impact of the work.
8. Metrics: Statement for the committee to track in order to assess the impact of the published standard over time to achieve the expected benefits.
9. Beneficiaries: Statement identifying and describing affected stakeholders and how they will each benefit from the proposal. E.g., list of beneficiaries.
   - Public policy holders.
   - Companies.
   - Consumers.
   - General public: Potential societal benefits include improved health, reduced pollution, increased employment, noiselessness, etc.

12.2.5 Process of Adoption of a National Standard on Biogas

An articulated synergy between these organizations will produce standards for biogas in Nigeria.

12.2.6 Steps in Adopting the Standard

The International Standard Organization (ISO) has the final say on any standard. The apex standard body in Nigeria is the SON. The ECN has the mandate for RE in Nigeria. An articulated synergy between these organizations will produce standards for RE in Nigeria. Figure 12.1 is the steps involved in adopting the biogas standard.

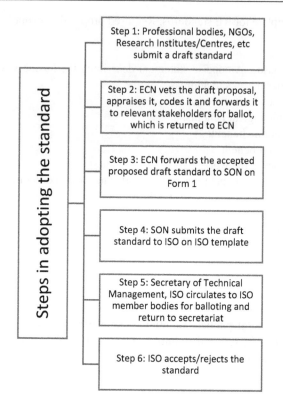

FIGURE 12.1   Steps in adopting a national standard on biogas.

A Biogas Curriculum for Nigerian Schools

13

A curriculum is a set of courses, coursework and their content, offered at a school or university.

13.1 BIOGAS COURSES

The proposed biogas courses are as follows:

BIG 01 – Introduction to biogas
BIG 02 – Production of biogas
BIG 03 – Biogas plants
BIG 04 – Design of biogas plants
BIG 05 – Construction of biogas plants
BIG 06 – Operating a biogas plant
BIG 07 – Biogas appliances
BIG 08 – Utilization of biogas
BIG 09 – Utilization of effluent from the plant

13.2 COURSE SYNOPSIS

A detailed description of the courses is as follows:

DOI: 10.1201/9781003241959-13

13.2.1 BIG 01 – Introduction to Biogas

Biogas: Definition; Explanation of biogas terms; Why biogas? Advantages and disadvantages of biogas. Detailed planning guide: Determining the energy demand; determining the biogas production; Balancing the gas production and demand by iteration–sample calculation.

Economic analysis and socio-economic evaluation: Procedures and target groups; Working time balance; Micro-economic analysis for the user-use of complex dynamic methods; Qualitative evaluation by the user; Macro-economic analysis and evaluation. Social acceptance and Dissemination: Determining factors of acceptance for biogas plants; Dissemination strategies; Implementing agencies; Artisan involvement; Training; Financing.

13.2.2 BIG 02 – Production of Biogas

Production: Anaerobic digestion; three-phase process; Feedstocks. Microorganisms responsible for biogas production: Bacteria; Fungi. Factors affecting production: Temperature; TS of slurry; Retention time; Loading rate of the slurry; pH of slurry; Carbon-to-nitrogen ratio of slurry; Toxicity. Improving biogas production: Mixed slurries; Pretreatment-heating, grinding, two-stage production, etc. Others: Biogas yield; Biogas quality; Production failure.

13.2.3 BIG 03 – Biogas Plants

Plant types: The floating-drum plant; The fixed dome plant; The balloon plant. Parts of the plants and their function: The digester well; The gasholder; The influent and effluent pits; The water jacket; The influent and effluent pipes. Sizes of plants: Small plants; Large plants. Pressure of plants: Low-pressure plants; Medium-pressure plants; High-pressure plants. Regional plants: Plants in cold regions; Plants in warm regions.

13.2.4 BIG 04 – Design of Biogas Plants

Sizing of plants: Scaling of the digester; Scaling of the gasholder; Gasholder/Digester ratio. Design of accessories: Gas pipes; Selection of valves; Selection of fittings.

13.2.5 BIG 05 – Construction of Biogas Plants

Pre-construction considerations: Masonry and mortar; Checklist for building a plant. Construction of plant: Inlet and outlet pits; Laying of inlet and outlet pipes; The mixing pit; Gasholder; Digester. Post-construction: Testing for water tightness of the digester; Testing for leakage in the gasholder; Testing for the system pressure.

13.2.6 BIG 06 – Operating Biogas Plants

Start-up of the plant: Seeding; Determining the loading rate; Slurry preparation. Plant operating pressure: Regulating the plant operating pressure; Measuring the plant pressure. Maintenance and Repair of plant: Troubleshooting a plant. Safety: Release of the first drumful of gas; Safety considerations.

13.2.7 BIG 07 – Biogas Appliances

Biogas stove: Types; Description and design; Testing. Incandescent lamp: Types; Description and design; Testing. Biogas IC engines: Description and design; Carburetion of air/biogas mixture.

13.2.8 BIG 08 – Utilization of Biogas

Heating: Operational considerations; Pressure loss at the stove. Lighting: Operational considerations; Pressure loss at the lamp. Electricity: Pressure loss at the IC engine; Gas quality factor; Operational considerations.

13.2.9 BIG 09 – Utilization of
Effluent from the Plant

Crop manure: Nutrient content of slurry; Soil fertilization; Treatment and handling of slurry for fertilization; Applying slurry to the soil. Animal feed: Treatment of slurry for animal feed; Compounding feed with treated slurry.

Appendices

APPENDIX 1: MEASUREMENT OF BIOGAS YIELD AND CONSUMPTION

The gas production from a biogas plant can be measured by use of a dry gas meter or by measuring the fill level of the gasholder. The floating drum in the water jacket on the digester well of a floating-drum plant is marked/calibrated such that the vertical rise and fall of the drum as a result of gas produced and gas consumed (Figure A1.1) can be read. The daily gas yield from the plant can be estimated from equation A1.1.

$$V_d = A_g \Delta h$$

$$V_d = \frac{\pi d^2}{4} \Delta h \qquad\qquad (A1.1)$$

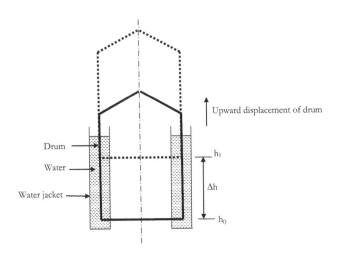

FIGURE A1.1 Upward displacement of the gasholder in the water jacket.

where

V_d is the gas yield from the biogas plant, m³
A_g is the area of the gasholder, m³
d is the diameter of the gasholder, m
Δh is the difference between h_0 and h_1, which is the upward displacement of the drum because of gas production, m

The gas consumed can be estimated from equation A1.2.

$$V_c = \frac{\pi d^2}{4} \Delta h \qquad\qquad\qquad (A1.2)$$

where

V_c is the volume of gas consumed, m³
d is the diameter of the gasholder, m
Δh is the difference between h_1 and h_0, which is the downward displacement of the drum because of gas removal from the drum (gas consumption), m³

APPENDIX 2: DISTRIBUTION OF BIOGAS PLANTS IN NIGERIA

TABLE A2.1 Biogas plants in Nigeria, distribution by states

S/NO.	LOCATION	SIZE (M³)	FEEDSTOCK	USE	TYPE OF PLANT	BUILT BY	DATE INSTALLED	STATUS
Federal Capital Territory, Abuja								
1	Federal Housing, Lugbe, Abuja	10.0	Sewage & food waste	Electricity & cooking	Fixed dome	ALI. Ltd (Engr. N. Ani)	2015	Functional
Anambra State								
2	Abattoir, Awka – Etiti market	10.0	Cow dung	Roasting of cow skin		NCERD (Staff)	2002	Not functional
Bayelsa State								
3	Sagbama, Rivers state	60.0	Human & Poultry waste	Electricity	Fixed dome with balloon storage system	ALI Int. Ltd (Engr. N. Ani)	2020	Functional
Benue State								
4	University of Agriculture, Makurdi	3.0	Cow dung	Training & research	Floating drum	Prof. I. N. Itodo	1999	Functional
5	HUDCO Qtr. North Bank, Makurdi	3.0	Human waste in soak-away	Cooking	Fixed dome	B. Beeka	2010	Functional

(Continued)

TABLE A2.1 (*Continued*)   Biogas plants in Nigeria, distribution by states

SINO.	LOCATION	SIZE (M³)	FEEDSTOCK	USE	TYPE OF PLANT	BUILT BY	DATE INSTALLED	STATUS
Borno State								
6	Private residence, Maiduguri	20.0	Cow dung	Cooking	Fixed dome		2000	Unknown
Cross River State								
7	Windsworth & University of Calabar	150.0	Human waste in soak-away	Cooking	Fixed dome	ALI Ltd (Engr. N. Ani)	2018	Functional
Delta State								
8	FCOE (Technical), Asaba	10.0	Cow dung	Heat for laboratory practical	Floating drum	Prof. J. I. Dioha	2011	Functional
9	FUPRE, Effurum	30.0	Cow dung	Drive a 5 kVA electric generator & research	Fixed dome	ALI. Ltd (Engr. N. Ani)	2020	Uncompleted
10	PIND Foundation, Warri	3.0	Kitchen waste	Cooking	Portable digester	ALI. Ltd (Engr. N. Ani)	2019	Functional

(*Continued*)

TABLE A2.1 (Continued) Biogas plants in Nigeria, distribution by states

S/NO.	LOCATION	SIZE (M³)	FEEDSTOCK	USE	TYPE OF PLANT	BUILT BY	DATE INSTALLED	STATUS
Edo State								
11	RRIN, Iyanomo, Benin-City	10.0	Rubber sheet effluent & cow dung	Heating	Floating drum		2014	Not functional
12	NCC, Enugu	150.0	Human waste in soak-away	Cooking	Fixed dome	ALI. Ltd (Engr. N. Ani)	2020	Functional
13	NCERD, UNN	10.0	Cow dung	Cooking & demonstration for electricity generation	Fixed dome	NCERD (Staff)	2002	Functional
14	Achalla village, Nsukka LGA	10.0	Cow dung	Frying garri		NCERD (Staff)	2004	Not functional
15	NCEEE, University of Benin	Two plants each of 10.0	Cow dung & water hyacinth	Cooking	Fixed dome 2019	NCERD (Staff)	2017	Functional
Enugu State								
16	Ugbo Nkwakwu village, Abakpa Nike	10.0	Cow dung	Frying garri		NCERD (Staff)	2001	Not functional

(Continued)

TABLE A2.1 (Continued) Biogas plants in Nigeria, distribution by states

S/NO.	LOCATION	SIZE (M³)	FEEDSTOCK	USE	TYPE OF PLANT	BUILT BY	DATE INSTALLED	STATUS
17	Guest House, NCERD, UNN	Five plants, each of 10.0	Human waste in soak-away	Cooking		NCERD (Staff)	2001	Not yet completed
Gombe State								
18	NCC, Gombe	100.0	Human waste in soak-away	Cooking	Fixed dome	ALI. Ltd (Engr. N. Ani)	2020	Functional
Imo State								
19	NCC, Owerri	10.0	Human waste in soak-away	Cooking	na	NCERD (Staff)	2005	Functional
Kaduna State								
20	NAERL, ABU	10.0	Cow dung	Heat for laboratory practical	na	na	1996	na
21	NCC, Zaria	50.0	Human waste in soak-away	Cooking	Fixed dome	na	1998	na
22	Ajarms Integrated Farms, Kaduna	30.0	Cooking	Fixed dome	na	2015	na
23	Science Secondary School, Zaria	20.0	Mixed livestock wastes	Heat for laboratory practical	Fixed dome	na	2012	na

(Continued)

TABLE A2.1 (Continued)   Biogas plants in Nigeria, distribution by states

S/NO.	LOCATION	SIZE (M³)	FEEDSTOCK	USE	TYPE OF PLANT	BUILT BY	DATE INSTALLED	STATUS
Katsina State								
24	GGASS, Shema Village, Dutse-Ma	30.0	Mixed livestock wastes	Cooking	Fixed dome	SERC (Staff)	2014	Functional Not in use because of non-availability of waste
25	NCERD, UYU, Katsina	30.0	Cow dung & other livestock wastes	Training & Research	Fixed dome	na	2015	na
Kebbi State								
26	Dalija Village	30.0	Cow dung	Heating & cooking	Fixed dome	SERC (Staff)	2006	Unknown
27	Nagari College, Birnin-Kebbi	30.0	Cow dung	Heating & cooking	Fixed dome	SERC (Staff)	2009	Functional Not in use because of non-availability of waste
Lagos State								
28	Abattoir, Oko-Oba, Agege, Lagos	10.0 & 20.0	Cow dung	Electricity	Fixed dome	ALI. Ltd (Engr. N. Ani)	2013	Not functional

(Continued)

TABLE A2.1 (Continued) Biogas plants in Nigeria, distribution by states

S/NO.	LOCATION	SIZE (M³)	FEEDSTOCK	USE	TYPE OF PLANT	BUILT BY	DATE INSTALLED	STATUS
29	Ojokoro Multi-purpose Farm, Lagos	20.0	Pig waste	Cooking	na	na	2001	na
30	NCC, Kirikiri, Lagos	200.0	Human waste in soak-away	Cooking	Fixed dome with balloon storage system	ALI. Ltd (Engr. N. Ani)	2017	Functional
31	Megamound Investment Company, Lekki - Lagos	200.0	Human waste in soak-away	Electricity	Fixed dome	ALI. Ltd (Engr. N. Ani)	2019	Completed Not yet in use
32	Jigbeko Farm Settlement, Gberibe – Ikorodu, Lagos	30.0	Pig waste	Electricity	Fixed dome	ALI. Ltd (Engr. N. Ani)	2019	Functional
33	Maranatha Vill Estate, Lekki – Lagos	20.0	Human waste in soak-away	Electricity	Fixed dome	ALI. Ltd (Engr. N. Ani)	2017	Not functional
Nassarawa State								
34	NCC, Lafia	100.0	Human waste in soak-away	Cooking	Fixed dome	ALI. Ltd (Engr. N. Ani)	2020	Functional

(Continued)

TABLE A2.1 (Continued) Biogas plants in Nigeria, distribution by states

SINO.	LOCATION	SIZE (M³)	FEEDSTOCK	USE	TYPE OF PLANT	BUILT BY	DATE INSTALLED	STATUS
Niger State								
35	Maryam Babangida Girls College, Minna	30.0	Cow dung	Heat for laboratory practical		SERC (Staff)	2010	Functional Not in use because of non-availability of waste
36	Kutunku Farms, Minna	140.0	Mixed livestock wastes	Cooking & electricity	Fixed dome	na	2013	na
Ogun State								
37	Mayflower School, Ikenne	20.0	Poultry droppings	Cooking	Fixed dome	SERC (Staff)	1994	No longer in existence. It is now a dumpsite
38	McNichole Plc., Papalanto	180.0	Poultry waste	Electricity	Fixed dome	ALI. Ltd (Engr. N. Ani)	2016	Not functional
Ondo State								
39	Federal College of Agriculture, Akure	na	Cow dung	Heating & lighting	na	na	na	na

(Continued)

TABLE A2.1 (*Continued*) Biogas plants in Nigeria, distribution by states

S/NO.	LOCATION	SIZE (M³)	FEEDSTOCK	USE	TYPE OF PLANT	BUILT BY	DATE INSTALLED	STATUS
Osun State								
40	Enpost farm, Ilesa	na	Livestock wastes	Heating & lighting	na	na	1982	na
Oyo State								
41	Safush garri processing farm, Alore – Moniya	40.0	Cassava peels & pig waste	Electricity	Fixed dome	ALI. Ltd (Engr. N. Ani)	2018	Functional
42	Federal Ministry of Agriculture, Okeho	20.0	Food waste	Heat for drying of crops	Fixed dome	ALI. Ltd (Engr. N. Ani)	2013	Not functional
Sokoto State								
43	SERC, UDUS (Old site)	30.0	Cow dung	Training	Fixed dome	SERC (Staff)	1985	Functional
44	SERC, UDUS (Old site 2)	30.0	Cow dung	Training & Research	Fixed dome	SERC (Staff)	1996	Functional
45	UDUS Secondary School	30.0	Cow dung	Heat for laboratory practical	Fixed dome	SERC (Staff)	na	Functional

(*Continued*)

TABLE A2.1 (Continued) Biogas plants in Nigeria, distribution by states

SINO.	LOCATION	SIZE (M³)	FEEDSTOCK	USE	TYPE OF PLANT	BUILT BY	DATE INSTALLED	STATUS
46	SERC, UDUS (Permanent site)	30.0	Cow dung	Training & Research	Fixed dome	SERC (Staff)	2003	Functional
47	Danjawa Solar Village, UDUS	20.0	Cow dung	Cooking & heating in the traditional ruler's house	Fixed dome	SERC (Staff)	2007	Functional. Not in use because of non-availability of waste
Zamfara State								
48	Government Science School, Gusau	20.0	Mixed livestock wastes	Heat for laboratory practical	Fixed dome	SERC (Staff)	2010	Functional. Not in use because of non-availability of waste

Source: Adapted and edited from Itodo et al. (2017).
na – not available.

APPENDIX 3: BIOGAS RESEARCHERS IN NIGERIA

A
Abubakar, A
Dr.
Biological Science Department
Botany Unit
Usmanu Danfodiyo University
Sokoto
Adeniran, A. Kamarudeen
Engr. Prof.
Department of Agricultural Engineering
Faculty of Engineering
University of Ilorin
Ilorin
Adeoti, Olusegun
Engr.
Department of Agricultural Engineering
Federal Polytechnic
Ado-Ekiti
Agbo, K. E
Dr.
National Centre for Energy Research & Development
University of Nigeria
Nsukka
Akinbami, John Felix Kayode
Dr.
Centre for Energy Research & Development
Obafemi Awolowo University
Ile-Ife
Akinwumi, I. O
Dr.
Technology Planning & Development Unit
Obafemi Awolowo University
Ile-Ife
Aliyu, H
Dr.
Sokoto Energy Research Centre
Usmanu Danfodiyo University

Sokoto
Ani, Nina
Engr. Mrs.
MD/CEO, Avenam Links Int. Ltd.
Suite F4, F & F Plaza
Adebisi Owoshogu Street
Off Cement Bus Stop
Ikeja-Lagos
Argungu, Garba M
Dr.
Department of Chemistry
Usmanu Danfodiyo University
Sokoto
Asere, A. Abraham
Engr. Prof.
Department of Mechanical Engineering
Lizade University
Ilaramokin
Ondo State
Atanu, Sunday Okpanachi
Engr.
Department of Agricultural Engineering
Federal Polytechnic
Bauchi
Awulu, John Okanagba
Engr. Dr.
Department of Agricultural & Environmental Engineering
College of Engineering
University of Agriculture
Makurdi
Ahmed, A
Mechanical Engineering Technology
Federal Polytechnic
Kaura Namoda
Zamfara State

B
Bala, E. J
Engr. Prof.
Director General & Chief Executive Officer
Energy Commission of Nigeria
Central District

Abuja
Bamgboye, Isaac
Engr. Prof.
Department of Agricultural & Environmental Engineering
Faculty of Technology
University of Ibadan
Ibadan
Buhari, M
Department of Physics
Kebbi State University of Science & Technology
Aliero
Kebbi state
Baki, Aliyu
Dr.
Usmanu Dafodiyo University
Sokoto

D
Dangogo, S. M
Prof.
Department of Chemistry
Usmanu Danfodiyo University
Sokoto
Dioha, J. I
Prof.
Renewable Energy Centre
Admiralty University of Nigeria
Ibusa
Delta State

E
Ekpekurede, A
Dr.
Department of Chemistry & Applied Chemistry
Usmanu Danfodiyo University
Sokoto
Eze, C. L
Prof.
Department of Environmental Management
River State University
Port Harcourt
Eze, I. John

Prof.
Department of Food Science & Technology
University of Nigeria
Nsukka
Ezeonu, F. C
Prof.
Department of Applied Biochemistry
Nnamdi Azikiwe University
Awka

G
Garba, Bashir
Prof.
Sokoto Energy Research Centre
Usmanu Danfodiyo University
Sokoto

I
Igoni, A. Hezekiah
Engr. Dr.
Department of Agricultural & Environmental Engineering
River State University
Port Harcourt
Igwilo, Kevin C
Dr.
Department of Petroleum Engineering
Federal University of Technology
Owerri
Ilori, M. O
Prof.
Technology Planning & Development Unit
Obafemi Awolowo University
Ile-Ife
Itodo, Isaac Nathaniel
Engr. Prof.
Department of Agricultural & Environmental Engineering
College of Engineering
University of Agriculture
Makurdi
Ibrahim, Y. Y.
Science Laboratory Technology
Federal Polytechnic

Kaura Namoda
Zamfara state

J
Jegede, Abiodun
Dr.
Obafemi Awolowo University
Ile-Ife

K
Kaankuka, Theresa Philip
Engr. Dr.
Department of Agricultural & Environmental Engineering
College of Engineering
University of Agriculture
Makurdi

L
Lucas, Emmanuel Babajide
Engr. Prof.
Department of Agricultural & Environmental Engineering
Faculty of Technology
University of Ibadan
Ibadan

M
Musa, Muazu
Dr.
Sokoto Energy Research Centre
Usmanu Danfodiyo University
Sokoto
Muhammad, I. B.
Ibrahim Shehu Shema Centre for Renewable Energy Researh
Umaru Musa Yar'dua University
Katsina

N
Ngumah, Chima
Dr.
Department of Microbiology
Federal University of Technology
Owerri

O

Ofoefule, A. U
Dr.
Department of Chemistry
University of Nigeria
Nsukka
Ogbulie, Nnama Jude-Anthony
Prof.
Department of Microbiology
Federal University of Technology
Owerri
Ogbuagu, J. O
Prof.
Department of Chemistry
Nnamdi Azikiwe University
Awka
Ojike, O
Dr.
Department of Agricultural & Bio-resources Engineering
University of Nigeria
Nsukka
Okaka, A. N. C
Prof. Mrs.
Department of Biochemistry
Nnamdi Azikiwe University
Awka
Okoro, Emeka Emmanuel
Dr.
Department of Petroleum Engineering
Covenant University
Ota – Lagos
Okwu, O. Modestus
Engr. Dr.
Department of Mechanical Engineering
College of Engineering & Technology
Federal University of Petroleum Resources
Effurum, Delta state
Orji, Justina Chibuogwu
Dr.
Department of Microbiology
Federal University of Technology
Owerri

Otanocha, Omonigho Benedict
Engr. Dr.
Department of Mechanical Engineering
College of Engineering & Technology
Federal University of Petroleum Resources
Effurum, Delta state

S
Sambo, Abubakar Sani
Engr. Prof. OON
Department of Mechanical Engineering
Usmanu Danfodiyo University
Sokoto
Samuel, O. David
Engr. Dr.
Department of Mechanical Engineering
College of Engineering & Technology
Federal University of Petroleum Resources
Effurum, Delta state
Sanni, E. Samuel
Dr.
Covenant University
Ota – Lagos
Shehu, S
Department of Physics
Umaru Musa Yar'dua University
Katsina

T
Tambuwal, A. Dahiru
Prof. Mrs.
Sokoto Energy Research Centre
Usmanu Danfodiyo University
Sokoto

U
Uba, Ahmed
Prof.
Department of Pure & Applied Chemistry
Usmanu Danfodiyo University
Sokoto
Udedi, S. E

Prof.
Department of Biochemistry
Nnamdi Azikiwe University
Awka
Ugwuishiwu, B. O
Engr. Dr.
Department of Agricultural & Bio-resources Engineering
University of Nigeria
Nsukka
Uzodinma, E. O
Dr.
National Centre for Energy Research & Development
University of Nigeria
Nsukka

Y
Yusuf, Paul Onalo
Engr.
Department of Agricultural & Environmental Engineering
College of Engineering
University of Agriculture
Makurdi

Z
Zuru, A. Abdullahi
Prof.
Department of Chemistry
Usmanu Danfodiyo University
Sokoto

References

Bari, S. (1996). Effect of carbon dioxide on the performance of biogas/diesel dual fuel engine: *Renewable Energy*, Vol. 9 (1–4): 1007–1010.

Bausa, M. (2011). *Studies on Balloon Storage of Biogas for Anywhere Use.* Unpublished Postgraduate Diploma project. Department of Agricultural and Environmental Engineering, University of Agriculture, Makurdi, Nigeria.

Becker, B. (2000). Calculating fuel consumption. *Boating Magazine* (14 Feb. 2000). Retrieved from https://www.boatingmag.com/calculating-fuel-consumption.

Bedoya, I., Cadavid, F., Saxena, S., Dibble, R., Aceves, S.M. and Flowers, D. (2012). A sequential chemical kinetics – CFD - chemical kinetics methodology to Predict HCCI combustion and main emissions. *SAE Technical Paper Series* 2012-01-1119. 10.4271/2012-01-1119.

Bolte, J.P., Hill, D.T. and Wood, T.H. (1986). Anaerobic digestion of screened waste liquids in suspended particle-attached growth reactors. *Transactions of ASAE*, Vol. 24 (20): 543–549.

ECN. (2005). Renewable energy resources, technology and markets. Renewable Energy Master Plan. Energy Commission of Nigeria, Abuja. pp. 3–4.

Heisler, M. (1981). Biogas filteration and storage. In: Waihyo, I. J. (2015). *Studies on the Storage of Biogas from a Floating Drum Plant. Bachelor of Engineering Project.* Department of Agricultural and Environmental Engineering, University of Agriculture, Makurdi, Nigeria. pp. 37–40.

Hill, D.T. and Bolte, J.P. (1986). Evaluation of SPAG fermenter treating liquid waste. *Transactions of ASAE*, Vol. 29 (6):1733–1738.

Itodo, I.N and Awulu, J.O. (1999). Effects of total solids concentration of poultry, cattle and piggery slurries on biogas yield. *Transactions of the ASAE*, Vol. 42(6): 1853–1855.

Itodo, I.N., Agyo, G.E. and Yusuf, P. (2007a). Performance evaluation of a biogas stove for cooking in Nigeria. *Journal of Energy in Southern Africa*, Vol. 18 (3): 14–18.

Itodo, I.N., Lucas, E.B. and Kucha, E.I. (1997a). The effect of using solar energy in the thermophilic digestion of poultry waste. *Journal of Engineering for International Development*, Vol. 3 (1): 15–21.

Itodo, I.N., Lucas, E.B. and Kucha, E.I. (1997b). The suitability of Obeche (Triplochiton Sceleroxylon) as media material in the anaerobic digestion of poultry waste. *Journal of Engineering for International Development*, Vol. 3 (1): 30–36.

Itodo, I. N., J. I. Dioha, P. A. Onwualu and Ojosu, J. (2017). Status of Renewable Energy in Nigeria. Energy Service Bulletin No. 1. Solar Energy Society of Nigeria.

Itodo, I. N., Yakubu, D.K. and Kaankuka, T.K. (2019). The effects of biogas fuel in an electric generator on greenhouse gas emissions, power output and fuel consumption. *Transactions of the American Society of Agricultural & Biological Engineers*. Vol. 62 (4): 951–958. ISSN 2151-0032. https://doi.org/10.1303/trans.13394. http://www.asabe.org.

Sasse, L. (1988). *Biogas Plants*. 2nd edition. Vieweg and Sohn, Braunschweig, Germany.

Sasse, L., Kellner, C. and Kimaro, A. (1991). *Improved Biogas Units for Developing Countries*. Vieweg and Sohn, Braunschweig, Germany.

Waihyo, J (2015). *Studies on the Storage of Biogas from a 3 m³ Floating Drum Plant. Unpublished Bachelor of Engineering Project*. Department of Agricultural and Environmental Engineering, University of Agriculture, Makurdi, Nigeria.

Werner, U., Stohr, U. and Hoes, N. (1989). *Biogas Plants in Animal Husbandries*. A Practical Guide. Vieweg and Sohn, Braunschweig, Germany.

Yusuf, P.O (2010). *Studies on Biogas Yield, Quality and Pressure as Affected by Retention Time and Weight on a 3 m³ Floating Drum Continuous-Flow Plant*. Unpublished Master of Engineering Thesis. Department of Agricultural and Environmental Engineering, University of Agriculture, Makurdi, Nigeria.

Index